Andreas Alfons

Simulation and Robust Statistics

Andreas Alfons

Simulation and Robust Statistics

Application to Laeken Indicators and Quality of Life Research

Südwestdeutscher Verlag für Hochschulschriften

Impressum/Imprint (nur für Deutschland/only for Germany)
Bibliografische Information der Deutschen Nationalbibliothek: Die Deutsche Nationalbibliothek verzeichnet diese Publikation in der Deutschen Nationalbibliografie; detaillierte bibliografische Daten sind im Internet über http://dnb.d-nb.de abrufbar.

Alle in diesem Buch genannten Marken und Produktnamen unterliegen warenzeichen-, marken- oder patentrechtlichem Schutz bzw. sind Warenzeichen oder eingetragene Warenzeichen der jeweiligen Inhaber. Die Wiedergabe von Marken, Produktnamen, Gebrauchsnamen, Handelsnamen, Warenbezeichnungen u.s.w. in diesem Werk berechtigt auch ohne besondere Kennzeichnung nicht zu der Annahme, dass solche Namen im Sinne der Warenzeichen- und Markenschutzgesetzgebung als frei zu betrachten wären und daher von jedermann benutzt werden dürften.

Verlag: Südwestdeutscher Verlag für Hochschulschriften GmbH & Co. KG
Dudweiler Landstr. 99, 66123 Saarbrücken, Deutschland
Telefon +49 681 37 20 271-1, Telefax +49 681 37 20 271-0
Email: info@svh-verlag.de

Zugl.: Wien, Technische Universität Wien, Dissertation, 2010

Herstellung in Deutschland:
Schaltungsdienst Lange o.H.G., Berlin
Books on Demand GmbH, Norderstedt
Reha GmbH, Saarbrücken
Amazon Distribution GmbH, Leipzig
ISBN: 978-3-8381-2706-4

Imprint (only for USA, GB)
Bibliographic information published by the Deutsche Nationalbibliothek: The Deutsche Nationalbibliothek lists this publication in the Deutsche Nationalbibliografie; detailed bibliographic data are available in the Internet at http://dnb.d-nb.de.

Any brand names and product names mentioned in this book are subject to trademark, brand or patent protection and are trademarks or registered trademarks of their respective holders. The use of brand names, product names, common names, trade names, product descriptions etc. even without a particular marking in this works is in no way to be construed to mean that such names may be regarded as unrestricted in respect of trademark and brand protection legislation and could thus be used by anyone.

Publisher: Südwestdeutscher Verlag für Hochschulschriften GmbH & Co. KG
Dudweiler Landstr. 99, 66123 Saarbrücken, Germany
Phone +49 681 37 20 271-1, Fax +49 681 37 20 271-0
Email: info@svh-verlag.de

Printed in the U.S.A.
Printed in the U.K. by (see last page)
ISBN: 978-3-8381-2706-4

Copyright © 2011 by the author and Südwestdeutscher Verlag für Hochschulschriften GmbH & Co. KG and licensors
All rights reserved. Saarbrücken 2011

Preface and Acknowledgments

First of all, I would like to thank my advisor Peter Filzmoser for his academic guidance and motivation. In addition, I am obliged to Matthias Templ for giving me the opportunity to do this research. Moreover, I am indebted to both of them, and of course to the other coauthors of my publications, Wolfgang E. Baaske, Josef Holzer, Stefan Kraft, Wolfgang Mader and Roland Wieser, for their contributions in our successful collaborations. To Josef Holzer, Stefan Kraft, Angelika Meraner and Stefan Zechner, with all of whom I shared an office at times, I am grateful for the friendly working environment. Furthermore, I would like to thank my family for always believing in me.

The research for this dissertation was partly funded by the European Union (represented by the European Commission) within the 7th Framework Programme for Research (Theme 8, Socio-Economic Sciences and Humanities, Project AMELI, Grant Agreement No. 217322), and by a grant of the Austrian Research Promotion Agency (FFG), Project ErfolgsVision (Ref. No. 813000/10345).

It should be noted that the original title of this dissertation is *"On statistical simulation and robust statistics with application to Laeken indicators and quality of life research"*. Due to the length of the title, it needed to be changed for this book. In addition, some colored graphics were replaced by figures using a greyscale. At the time this dissertation was finished, Chapter 2 had been accepted for publication in the *Journal of Statistical Software*, and Chapter 8 had been accepted for publication in the journal *Statistical Methods & Applications*. Both chapters are now published (Alfons et al. 2010c, 2011a), therefore these references have been updated. In addition, Chapter 5 is a revised version of a paper submitted to the journal *Statistical Methods & Applications*. The final version of the paper is now accepted for publication and is available from the journal's website (Alfons et al. 2011b).

Andreas Alfons

Abstract

Due to the complexity of modern statistical methods, in particular in robust statistics, obtaining analytical results about their properties is often virtually impossible. Consequently, simulation studies are widely used by statisticians to gain insight into the quality of developed methods. In addition, research projects commonly involve many scientists, often from different institutions, each focusing on different aspects of the project. Hence precise guidelines regarding the design of simulation studies are necessary in order to draw meaningful conclusions. As a remedy, a general framework for statistical simulation designed to simplify obtaining comparable results in collaborative research projects has been implemented in the R package **simFrame**.

Simulation studies in survey statistics are typically performed by repeatedly drawing samples from a finite population. However, real population data are only in exceptions available to researchers. Therefore, suitable population data need to be generated synthetically. The simulated data need to be as realistic as possible, while at the same time ensuring data confidentiality. A method for generating close-to-reality population data for complex household surveys has thus been developed and implemented in the R package **simPopulation**. Furthermore, data confidentiality issues are analyzed using several different worst case scenarios.

The Laeken indicators are a set of indicators defined by the European Union for measuring poverty and social cohesion in Europe. However, some of these indicators are highly influenced by outliers in the upper tail of the income distribution. In order to reduce the influence of outlying observations, the use of robust Pareto tail modeling is investigated in a simulation setting. Selected Laeken indicators and methods for Pareto tail modeling have been implemented in the R package **laeken**.

Statistical models in the social sciences need to be limited to a small number of explanatory variables with low interdependencies for better interpretability. To achieve these goals, a robust model selection method, combined with a strategy to reduce the number of selected predictor variables to a necessary minimum, has been developed. In addition, the proposed method is applied to obtain responsible factors describing the cognition of quality of life in smaller municipalities.

Kurzzusammenfassung

Durch die Komplexität moderner statistischer Methoden, insbesondere in der robusten Statistik, ist es oftmals nahezu unmöglich, analytische Resultate über ihre Eigenschaften zu erzielen. Folglich ist der Einsatz von Simulationsstudien um einen Einblick in die Qualität der entwickelten Methoden zu gewinnen unter Statistikern weit verbreitet. Des Weiteren sind an Forschungsprojekten üblicherweise viele Wissenschafter beteiligt, oftmals von verschiedenen Institutionen, von denen sich jeder auf andere Aufgaben innerhalb des Projekts konzentriert. Deswegen sind genaue Richtlinien bezüglich des Designs der Simulationsstudien notwendig, um aussagekräftige Schlussfolgerungen ziehen zu können. Als Abhilfe wurde ein Framework für statistische Simulation entworfen und in dem R Paket **simFrame** implementiert, welches es in gemeinschaftlichen Forschungsprojekten erleichtert, vergleichbare Resultate zu erlangen.

Simulationsstudien in der offiziellen Statistik werden üblicherweise durchgeführt, indem wiederholt Stichproben aus einer endlichen Grundgesamtheit gezogen werden. Allerdings stehen Forschern echte Populationsdaten nur in Ausnahmefällen zur Verfügung, daher muss eine geeignete Grundgesamtheit künstlich erzeugt werden. Die simulierten Daten müssen so realistisch wie möglich sein, aber zugleich darf die statistische Geheimhaltung nicht verletzt sein. Dementsprechend wurde eine Methode zur Erzeugung von realitätsnahen Populationsdaten entwickelt und in dem R Paket **simPopulation** implementiert. Zusätzlich wird die statistische Geheimhaltung anhand verschiedener Worst Case Szenarien untersucht.

Die sogenannten Laeken Indikatoren sind eine Reihe von Indikatoren, die von der Europäischen Union zusammengestellt wurden, um Armutsgefährdung und sozialen Zusammenhalt innerhalb Europas zu messen. Jedoch sind einige dieser Indikatoren stark von Ausreißern am oberen Ende der Einkommensverteilung beeinflusst. Um den Einfluss solcher Ausreißer zu verringern, wird die robuste Modellierung des oberen Endes der Verteilung durch eine Pareto Verteilung mittels Simulationen untersucht. Ausgewählte Laeken Indikatoren und Methoden zur Modellierung einer Pareto Verteilung wurden in dem R Paket **laeken** implementiert.

Statistische Modelle in den Sozialwissenschaften müssen auf eine sehr kleine Anzahl von erklärenden Variablen mit geringen gegenseitigen Abhängigkeiten beschränkt sein, um eine bessere Interpretierbarkeit zu gewährleisten. Um diese Ziele zu erreichen, wurde ein Verfahren entwickelt, das robuste Modellselektion mit einer Strategie zur Reduktion der Anzahl an ausgewählten Variablen auf ein nötiges Minimum kombiniert. Zudem wird das entwickelte Verfahren angewendet, um jene hauptverantwortlichen Faktoren zu finden, welche die Wahrnehmung von Lebensqualität in kleineren Gemeinden erklären.

Contents

Preface and Acknowledgments	i
Abstract	iii
Kurzzusammenfassung	v
List of Tables	xi
List of Figures	xii

1 Introduction   **1**
- 1.1 Project AMELI 2
  - 1.1.1 Selected Laeken indicators 3
- 1.2 Project ErfolgsVision 4
- 1.3 Robust statistics 5
  - 1.3.1 Breakdown point 6
  - 1.3.2 Influence function 7
  - 1.3.3 Example: Covariance matrix estimation 7
  - 1.3.4 Outliers in survey statistics 8
- 1.4 Statistical simulation 9
  - 1.4.1 Random number generation 9
  - 1.4.2 General design of simulation studies in survey statistics 10
  - 1.4.3 Finite population sampling and weighting 12
  - 1.4.4 Contamination models 13
  - 1.4.5 Missing data models 14
  - 1.4.6 Parallel computing 17
- 1.5 Outline of the remaining chapters 18
  - 1.5.1 Overview of the remaining chapters 19

2 An object-oriented framework for statistical simulation   **21**
- 2.1 Introduction 22

CONTENTS

- 2.2 Object-oriented programming and S4 23
- 2.3 Design of the framework 24
  - 2.3.1 UML class diagram 26
  - 2.3.2 Naming conventions 26
  - 2.3.3 Accessor and mutator methods 28
- 2.4 Implementation 28
  - 2.4.1 Data handling 29
  - 2.4.2 Sampling 30
  - 2.4.3 Contamination 34
  - 2.4.4 Insertion of missing values 36
  - 2.4.5 Running simulations 38
  - 2.4.6 Visualization 42
- 2.5 Parallel computing 42
- 2.6 Using the framework 43
  - 2.6.1 Design-based simulation 44
  - 2.6.2 Model-based simulation 46
  - 2.6.3 Parallel computing 49
- 2.7 Extending the framework 51
  - 2.7.1 Model-based data 52
  - 2.7.2 Sampling 52
  - 2.7.3 Contamination 54
  - 2.7.4 Insertion of missing values 55
- 2.8 Conclusions and outlook 55

3 Applications of statistical simulation in the case of EU-SILC   57
- 3.1 Introduction 57
- 3.2 Application of different simulation designs to EU-SILC 58
  - 3.2.1 Basic simulation design 59
  - 3.2.2 Using stratified sampling 60
  - 3.2.3 Adding contamination 61
  - 3.2.4 Performing simulations separately on different domains 63
  - 3.2.5 Using multiple contamination levels 65
  - 3.2.6 Inserting missing values 67
  - 3.2.7 Parallel computing 69
- 3.3 Conclusions 72

4 Contamination models in the R package simFrame   73
- 4.1 Introduction 73
- 4.2 Contamination models in **simFrame** 74

	4.3	Example: Outlier detection	75
	4.4	Conclusions	77

5 Simulation of close-to-reality population data   79
- 5.1 Introduction . . . 80
- 5.2 Simulation of synthetic populations . . . 81
  - 5.2.1 Setup of the household structure . . . 83
  - 5.2.2 Simulation of categorical variables . . . 83
  - 5.2.3 Simulation of continuous variables . . . 85
  - 5.2.4 Splitting continuous variables into components . . . 88
  - 5.2.5 Software . . . 89
- 5.3 Application to EU-SILC . . . 90
  - 5.3.1 Diagnostic plots for a single simulation . . . 90
  - 5.3.2 Average results from multiple simulations . . . 97
  - 5.3.3 Influence of sample size and sampling design . . . 99
- 5.4 Conclusions . . . 101

6 Disclosure risk of synthetic population data   105
- 6.1 Introduction . . . 106
- 6.2 Generation of synthetic population data . . . 107
- 6.3 Synthetic EU-SILC population data . . . 108
- 6.4 A global disclosure risk measure for survey data . . . 110
- 6.5 Confidentiality of synthetic population data . . . 111
- 6.6 Disclosure scenarios for synthetic population data . . . 112
  - 6.6.1 Scenario 1 . . . 113
  - 6.6.2 Scenario 2 . . . 114
  - 6.6.3 Scenario 3 . . . 114
  - 6.6.4 Scenario 4 . . . 114
  - 6.6.5 Scenario 5 . . . 115
- 6.7 Results . . . 115
- 6.8 Conclusions . . . 116

7 A comparison of robust methods for Pareto tail modeling   119
- 7.1 Introduction . . . 119
- 7.2 Selected Laeken indicators . . . 120
  - 7.2.1 Quintile share ratio . . . 120
  - 7.2.2 Gini coefficient . . . 120
- 7.3 Pareto tail modeling . . . 121
  - 7.3.1 Hill estimator . . . 122

		7.3.2	Weighted maximum likelihood (WML) estimator	122
		7.3.3	Partial density component (PDC) estimator	123
	7.4	Simulation study		123
	7.5	Conclusions and outlook		124

8 Robust variable selection — 127

- 8.1 Introduction — 128
- 8.2 Context-sensitive model selection — 129
  - 8.2.1 Description of the algorithm — 130
  - 8.2.2 Summary of the algorithm — 132
  - 8.2.3 Diagnostics — 133
  - 8.2.4 Implementation — 134
- 8.3 Example: Driving factors behind quality of life — 134
  - 8.3.1 Results — 134
  - 8.3.2 CPU times — 139
- 8.4 Simulations — 139
- 8.5 Conclusions and discussion — 143

References 145

Index 158

List of Tables

5.1	Variables selected for the simulation of the Austrian EU-SILC population data.	91
5.2	Categorized variables created for use as predictors during the simulation.	92
5.3	Pairwise contingency coefficients of the categorical variables.	98
5.4	Evaluation of personal net income. Values from the sample data are compared to average results from 100 simulated populations.	99
5.5	Analysis of empty cells in the contingency tables of the categorical variables.	100
5.6	Pairwise contingency coefficients of the categorical variables for the initial population, as well as average results from 250 simulated populations for each of the four sampling scenarios.	102
5.7	Evaluation of personal net income. Values from the initial population are compared to average results from 250 simulated populations for each of the four sampling scenarios.	103
6.1	Variables of the synthetic EU-SILC population data.	109
6.2	Results for Scenarios 1-5.	116
8.1	Explanation of important variables.	135
8.2	MM-regression results for the RCS model for quality of life.	136
8.3	MM-regression results for the B-RLARS model for quality of life.	136
8.4	Average results from 100 simulation runs with contamination level $\varepsilon = 0.1$.	142

LIST OF TABLES

List of Figures

1.1	Lorenz curve.	5
1.2	Covariance matrix estimation in the two-dimensional case using classical and robust methods.	8
1.3	Outline of the most realistic design for simulation studies in survey statistics.	11
1.4	Outline of a more practical approach for the simulation studies in survey statistics.	12
1.5	Simulated bivariate data set with high correlation.	15
1.6	Simulated bivariate data set without correlation.	16
2.1	Slightly simplified UML class diagram of **simFrame**.	27
2.2	Simulation results from the simple illustrative example.	43
2.3	Default plot of results from a simulation study with one contamination level and different domains	46
2.4	*Left*: Default plot of results from a simulation study with one missing value rate. *Right*: Kernel density plots of the simulation results.	49
2.5	*Left*: Default plot of results from a simulation study with multiple missing value rates. *Right*: Kernel density plots of the simulation results for a specified missing value rate.	52
2.6	*Top:* Code skeleton for a user-defined data generation method. *Bottom:* Code skeleton for extending model-based data generation.	53
2.7	*Top:* User-defined function for Poisson sampling. *Bottom:* Code skeleton for user-defined setup of multiple samples.	53
2.8	Code skeleton for user-defined contamination.	54
2.9	Code skeleton for user-defined insertion of missing values.	55
3.1	Simulation results for the basic simulation design.	60
3.2	Simulation results for the simulation design with stratified sampling.	62
3.3	Simulation results for the simulation design with stratified sampling and contamination.	63

LIST OF FIGURES

3.4 Simulation results for the simulation design with stratified sampling, contamination and performing the simulations separately for each gender. 65

3.5 Simulation results for the simulation design with stratified sampling, multiple contamination levels and performing the simulations separately for each gender. 67

3.6 Simulation results for the simulation design with stratified sampling, multiple contamination levels, multiple missing value rates and performing the simulations separately for each gender. 69

3.7 Simulation results obtained by parallel computing for the simulation design with stratified sampling, multiple contamination levels, multiple missing value rates and performing the simulations separately for each gender. 71

4.1 Average proportions of false negatives (*left*) and false positives (*right*). . . . 76

5.1 *Top*: Mosaic plots of gender, region and household size. *Bottom*: Mosaic plots of gender, economic status and citizenship. 93

5.2 *Left*: Cumulative distribution functions of personal net income. *Right:* Box plots of personal net income. 94

5.3 Box plots of personal net income split by gender (*top left*), citizenship (*top right*), region (*bottom left*) and economic status (*bottom right*). 95

5.4 Box plots of the income components. 96

6.1 Mosaic plots of gender, region and household size of the Austrian EU-SILC sample from 2006 and the resulting synthetic population. 109

6.2 Personal net income in the Austrian EU-SILC sample from 2006 and the resulting synthetic population. 110

7.1 Simulation results for the quintile share ratio *(left)* and the Gini coefficient *(right)* without contamination. 125

7.2 Simulation results for the quintile share ratio *(left)* and the Gini coefficient *(right)* with 0.25% contamination. 125

8.1 Learning curve for the B-RLARS sequence (*left*). Densities of the RTMSEP (*right*). 137

8.2 Dendrogram of the initial B-RLARS sequence of candidate predictors for quality of life. 138

Chapter 1

Introduction

This dissertation has been written in the course of two research projects: the project AMELI (*Advanced Methodology of the European Laeken Indicators*), funded by the European Union within the 7th Framework Programme for Research (FP7), and the project *ErfolgsVision*, funded by the Austrian Research Promotion Agency (FFG). Both projects are based on applications of statistics in socioeconomics and are focused on related topics. While the aim of AMELI is to improve the estimation of European Union indicators of risk-of-poverty and social cohesion, ErfolgsVision tries to explain success factors of smaller municipalities. In a way, AMELI is directed at monitoring quality of life from a monetary perspective, whereas ErfolgsVision is interested in the cognition of quality of life. Moreover, robustness against outliers in the data is an important issue in both research projects, and the developed methods are evaluated by means of simulation studies. These two topics, statistical simulation and robust statistics, therefore are the main focus of this thesis.

The remaining chapters constitute a collection of papers, almost all of which are already published or have been accepted for publication (see Section 1.5.1). A broad range of topics is treated these papers: (*i*) development of a software framework for simulation studies in statistics, (*ii*) simulation of synthetic population data using statistical models and evaluation of the generated data with respect to data confidentiality issues, (*iii*) robust modeling of the upper tail of income data with a Pareto distribution in order to reduce the influence of outlying observations on inequality indicators, and (*iv*) robust model selection for applications in the social sciences. Even though the range of these topics is quite broad, they are closely linked together as outlined in the following.

Modern statistical methods, in particular in robust statistics, are highly complex and are thus frequently evaluated by simulation. When a research project is based on extensive simulation studies, precise guidelines for the simulations are required. Otherwise the results obtained different researchers may not be comparable, which in turn makes it impossible to draw meaningful conclusions from the project. Hence the developed framework for statistical simulation is designed to facilitate the coordination of simulation studies in such research

1. Introduction

projects. Special emphasis is thereby given to applications in survey statistics and robust estimation.

In survey statistics, simulation studies are typically performed by repeatedly drawing samples from a finite population (see Section 1.4.2). Nevertheless, suitable population data need to be synthetically generated in practice, since real population data are only in exceptions available to researchers. If such synthetic data are released to the public, which is important to allow for reproducibility of the simulation results by other scientists, data confidentiality must not be violated. With the developed methodology, it is possible to simulate population data for complex household surveys that are both of high quality and confidential.

Based on such synthetic population data, the estimation of social inclusion indicators is evaluated in an application of the proposed simulation framework. As mentioned above, Pareto distributions are fitted to the upper tail of income data by robust estimation in order to reduce the influence of outlying observations on inequality indicators.

Since the distribution of income is related to quality of life, it is also of interest to find statistical models that explain notions such as the cognition of quality of life. For better interpretability in the social sciences, the models need to be limited to a small number of explanatory variables with low interdependencies. Therefore, a robust model selection method, combined with a strategy to reduce the number of selected predictor variables to a necessary minimum, has been developed.

It is also important to note that all software developed for this thesis is freely available for the open-source statistical environment R (R Development Core Team 2010). Availability of open-source software is highly important in research so that scientists are able to use or even modify the latest state-of-the-art methods for their own work.

The rest of this introduction is organized as follows. In Section 1.1 and 1.2, the two research projects mentioned above are described in more detail. Section 1.3 gives a brief introduction on robust statistics. Basic concepts of statistical simulation are then discussed in Section 1.4. Finally, Section 1.5 gives an overview of the remaining chapters of this dissertation.

1.1 Project AMELI

In order to monitor risk-of-poverty and social cohesion in Europe for policy analysis purposes, the European Union introduced a set of indicators called the *Laeken* indicators. The EU-SILC (*European Union Statistics on Income and Living Conditions*) survey thereby serves as data basis for most of these indicators. EU-SILC is an annual panel survey conducted in European Union member states and other European countries. As indicated by its name, EU-SILC contains detailed information about the income of the sampled households. Most notably, the total disposable household income is equivalized with respect to the number of household members, and each person in the same household is assigned the same *equivalized disposable income* (for details on its computation, see EU-SILC 2004).

The project acronym AMELI stands for *Advanced Methodology of the European Laeken Indicators*. The aim of the project is not only to improve the estimation of the Laeken indicators directly, e.g., by applying robust methods, but also by enhancing the data quality with, e.g., appropriate imputation or outlier detection methods. Extensive simulation studies are thereby used to evaluate the developed methodology. Since many partners from different institutions across Europe are involved in the AMELI project, the simulation studies need to be well coordinated.

1.1.1 Selected Laeken indicators

Concerning robustness, the inequality indicators *quintile share ratio* (QSR) and *Gini coefficient* are of particular interest (see Chapter 7). However, due to its importance, the *at-risk-of-poverty rate* (ARPR) is mentioned here, too. The complete list of Laeken indicators based on EU-SILC along with their definitions can be found in EU-SILC (2004).

For the following definitions, let $\boldsymbol{x} := (x_1, \ldots, x_n)'$ be the equivalized disposable income with $x_1 \leq \ldots \leq x_n$ and let $\boldsymbol{w} := (w_i, \ldots, w_n)'$ be the corresponding personal sample weights, where n denotes the number of observations. Furthermore, let q_α with $0 \leq \alpha \leq 1$ denote the weighted α quantile of $\boldsymbol{x}$ with weights $\boldsymbol{w}$, and define the following index sets for a certain threshold t:

$$I_{<t} := \{i \in \{1, \ldots, n\} : x_i < t\}, \tag{1.1}$$

$$I_{\leq t} := \{i \in \{1, \ldots, n\} : x_i \leq t\}, \tag{1.2}$$

$$I_{>t} := \{i \in \{1, \ldots, n\} : x_i > t\}. \tag{1.3}$$

At-risk-of-poverty rate (ARPR)

For the definition of the at-risk-of-poverty rate (ARPR), the *at-risk-of-poverty threshold* (ARPT) needs to be introduced first, which is defined as 60% of the median equivalized disposable income:

$$ARPT := 0.6 \cdot q_{0.5}. \tag{1.4}$$

Using this definition of the at-risk-of-poverty threshold, the at-risk-of-poverty rate is defined as the percentage of persons (over the total population) with an equivalised disposable income below the at-risk-of-poverty threshold (EU-SILC 2004). In mathematical terms, it is estimated by

$$ARPR := 100 \cdot \frac{\sum_{i \in I_{<ARPT}} w_i}{\sum_{i=1}^n w_i}, \tag{1.5}$$

1. Introduction

where $I_{<ARPT}$ is an index set as defined in (1.1).

Quintile share ratio (QSR)

The income quintile share ratio (QSR) is defined as the ratio of the sum of equivalized disposable income received by the 20% of the population with the highest equivalized disposable income to that received by the 20% of the population with the lowest equivalized disposable income (EU-SILC 2004). Using index sets $I_{\leq q_{0.2}}$ and $I_{>q_{0.8}}$ as defined in (1.2) and (1.3), respectively, the quintile share ratio is estimated by

$$QSR := \frac{\sum_{i \in I_{>q_{0.8}}} w_i x_i}{\sum_{i \in I_{\leq q_{0.2}}} w_i x_i}. \tag{1.6}$$

Gini coefficient

The Gini coefficient is defined as the relationship of cumulative shares of the population arranged according to the level of equivalized disposable income, to the cumulative share of the equivalized total disposable income received by them (EU-SILC 2004). Mathematically speaking, the Gini coefficient is estimated by

$$Gini := 100 \left[\frac{2 \sum_{i=1}^{n} \left(w_i x_i \sum_{j=1}^{i} w_j \right) - \sum_{i=1}^{n} w_i^2 x_i}{\left(\sum_{i=1}^{n} w_i \right) \sum_{i=1}^{n} (w_i x_i)} - 1 \right]. \tag{1.7}$$

For a visual explanation of the Gini coefficient, consider the *Lorenz curve* (Lorenz 1905). It is obtained by plotting the cumulative share of the population against the cumulative share of income after sorting the observations according to income in ascending order. An example is shown in Figure 1.1. If the income would be distributed in perfect equality, i.e., each person receives the same income, the Lorenz curve would be a straight line from $(0,0)$ to $(1,1)$. In any case, the Gini coefficient is given by

$$Gini = 100 \cdot 2A, \tag{1.8}$$

where A denotes the area between the Lorenz curve and the line of perfect equality.

1.2 Project ErfolgsVision

The research project *ErfolgsVision* was a cooperation with the regional developer *SPES Academy* and the applied social research center *STUDIA-Schlierbach*. The aim of the project was to find statistical models for success factors of smaller municipalities, in particular for the cognition of quality of life. An important requirement was that the resulting models

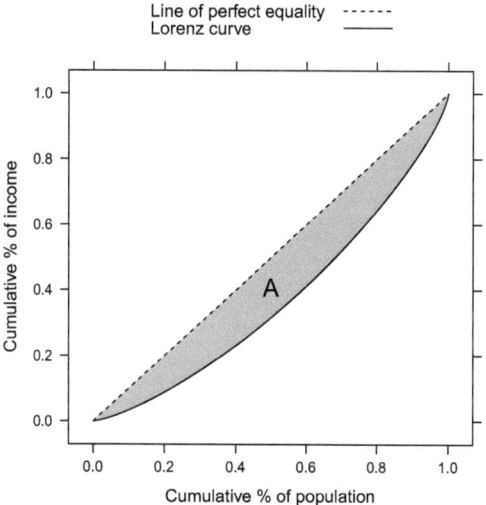

Figure 1.1: Lorenz curve.

should be highly interpretable in the context of the social sciences, therefore they should contain only very few key predictor variables and the dependencies between those predictors should be low.

In order to be able to conduct this research, a large database containing socioeconomic data from 60 communities in Austria and Germany together with data from official statistical institutes about these communities has been built. The socioeconomic data were thereby collected from 18 748 questionnaires filled out by inhabitants of the involved municipalities, which corresponds to an average of 312 questionnaires per municipality.

1.3 Robust statistics

Statistical methods frequently make some assumptions about the data such as a normal distribution. The fundamental principle of robust statistics is to allow certain parts of the data to deviate from the underlying assumptions. The parts of the data that deviate from the main part are thereby referred to as *contamination* and the respective observations as *outliers*. Consequently, in robust statistics the distribution F of such contaminated data is typically modeled as a mixture of distributions

$$F = (1-\varepsilon)G + \varepsilon H, \qquad (1.9)$$

where ε denotes the *contamination level*, G is the distribution of the main part of the data and H is the distribution of the contamination (e.g., Maronna et al. 2006).

The aim of robust statistical methods is to obtain estimates that are representative for the majority of the data. Note that robust methods therefore are in general less efficient than their classical counterparts, since they do not use all information from all observations, i.e., by leaving out observations, or better, by downweighting suspicious observations. In practice, it is thus often important to find a compromise between robustness and efficiency.

1.3.1 Breakdown point

Classical methods such as maximum likelihood estimators use information from all observed data points and are thus highly efficient if the underlying assumptions hold. However, if there are data points that deviate from these assumptions, the obtained results may not reflect the majority of the data anymore or may even be completely arbitrary. As an example, consider the sample mean $\bar{x}_n := \frac{1}{n}\sum_{i=1}^{n} x_i$, where n is the number of observations and $\boldsymbol{x} = (x_1 \ldots, x_n)'$ denotes the observed vales. If only a single observation x_k with $k \in \{1,\ldots,n\}$ fixed is replaced by some other value x'_k, then $\bar{x}_n$ can be moved arbitrarily far away from its original value by moving x'_k away from the rest of the data. To formalize this issue, Hampel (1971) introduced the notion of the *breakdown point* of an estimator, which has been further refined in the literature on robust statistics.

Let $\hat{\theta}_n = \hat{\theta}_n(\boldsymbol{X})$ be an estimate for the parameter θ given a sample $\boldsymbol{X}$ with n observations, and let the range of θ be denoted by the set Θ. Donoho and Huber (1983) defined the *replacement finite-sample breakdown point* $\varepsilon_n^*(\hat{\theta}_n, \boldsymbol{X})$ as the largest proportion of data points that can be replaced by outliers such that $\hat{\theta}_n$ remains bounded and also bounded away from the boundary $\partial\Theta$ of the parameter range Θ. For a more formal definition, let $\mathcal{X}_m$ be the set of all data sets with n observations that result from arbitrarily replacing m observations in $\boldsymbol{X}$. Then the replacement finite-sample breakdown point can be written as

$$\varepsilon_n^*(\hat{\theta}_n, \boldsymbol{X}) := \max_{0\leq m \leq n} \left\{ \frac{m}{n} : \hat{\theta}_n(\boldsymbol{Y}) \text{ bounded and bounded away from } \partial\Theta \ \forall \boldsymbol{Y} \in \mathcal{X}_m \right\}. \quad (1.10)$$

In addition, the *asymptotic breakdown point* ε^* is given by $\lim_{n\to\infty} \varepsilon_n^*$, which exists in most cases of interest (see Maronna et al. 2006). Thus the sample mean has an asymptotic breakdown point of 0, as a single outlier can distort the result to an arbitrary extent. It should be noted that $\varepsilon^* \leq \frac{1}{2}$ holds for any estimator, as there must be more typical than atypical data points. For further details on the breakdown point, including a formal proof of the last statement, the reader is referred to Maronna et al. (2006).

1.3.2 Influence function

Another important concept in robust statistics is the *influence function* of an estimator (Hampel 1974). As the name suggests, it is a measure for the influence of a small fraction of identical outliers on the asymptotic behavior of an estimator. In order for an estimator to be robust, its influence function must be bounded.

Let $\hat{\theta}_n = \hat{\theta}_n(\boldsymbol{X})$ again be an estimate for the parameter θ given a sample $\boldsymbol{X}$ with n i.i.d. observations coming from a distribution F. When $n \to \infty$, there exists a value $\hat{\theta}_\infty = \hat{\theta}_\infty(F)$ in most cases of practical interest such that $\hat{\theta}_n \to_p \hat{\theta}_\infty$, which is called the *asymptotic value* of the estimate at F (see Maronna et al. 2006). To continue the example from the previous section, the asymptotic value of the sample mean $\bar{x}_n$ is given by the distribution mean $\mathbb{E}_F(x)$. In any case, the influence function of an estimator $\hat{\theta}$ at distribution F is defined as

$$IF(x, \hat{\theta}, F) := \lim_{\varepsilon \to 0+} \frac{\hat{\theta}_\infty\left((1-\varepsilon)F + \varepsilon\delta_x\right) - \hat{\theta}_\infty(F)}{\varepsilon}, \tag{1.11}$$

where δ_x is a probability distribution that assigns mass 1 to point x. A more detailed discussion on the influence function can be found in Maronna et al. (2006).

1.3.3 Example: Covariance matrix estimation

As a further example for the necessity of robust statistics, consider covariance matrix estimation in the two-dimensional case. In this motivational example, the classical sample covariance matrix is compared to the robust *minimum covariance determinant* (MCD) estimator (Rousseeuw and Van Driessen 1999) using simulated bivariate data. For a given fraction α of data points, $\frac{1}{2} \leq \alpha \leq 1$, the MCD seeks the subset of points for which the determinant of the covariance matrix is minimal. The mean and covariance matrix of this subset then constitute robust estimates of location and scatter with an asymptotic breakdown point of $1 - \alpha$. Improved estimates are obtained by applying a finite sample correction factor and an asymptotic consistency factor. In practice, a suitable value for α needs to be chosen. Setting $\alpha = \frac{1}{2}$ leads to the maximum breakdown point any estimator can attain. If the proportion of outliers in the data is assumed to be smaller, increasing the value of α results in higher efficiency.

Figure 1.2 show data from a bivariate normal distribution with a group of outliers that are drawn from a normal distribution with a shifted mean and a deflated covariance matrix. In addition, 97.5% tolerance ellipses for the classical covariance matrix and the MCD with $\alpha = \frac{3}{4}$ are displayed. Clearly, the shape of the classical covariance matrix is distorted by the outlier group, while the MCD represents the covariance structure of the majority of the data very well.

1. Introduction

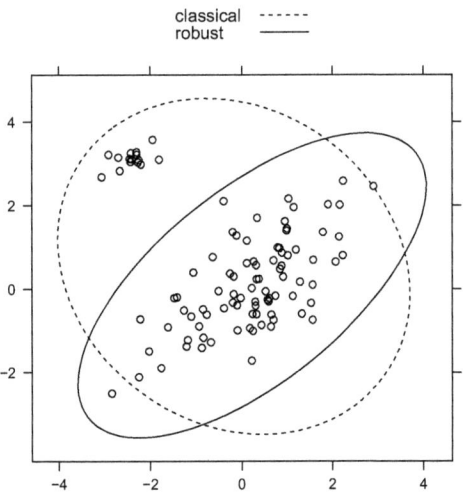

Figure 1.2: Covariance matrix estimation in the two-dimensional case using classical and robust methods.

1.3.4 Outliers in survey statistics

In survey statistics, Chambers (1986) introduced the notion of *representative* and *non-representative* outliers. When samples are drawn from a finite population, each sampled observation is assigned a weight and therefore represents a number of observations in the population. Unless a basic sampling procedure such as simple random sampling is used, the weights are in general not equal for all sampled observations and need to be considered when quantities of interest such as indicators are estimated, otherwise the obtained estimates will be biased.

Representative outliers in a sample are observations whose values have been recorded correctly and cannot be regarded as unique in the population. Therefore, these observations contain some relevant information for estimating quantities of interest. Nonrepresentative outliers, on the other hand, contain values that are either incorrect or unique to the specific population element. In the first case, they need to be corrected in the data editing process. If they are correct but can be considered unique in the population, they need to be downweighted for the estimation of quantities of interest in order not to corrupt the estimates.

1.4 Statistical simulation

In this section, some general principles of statistical simulation with special emphasis on survey statistics are discussed. *Monte Carlo* simulation is widely used in statistics as a computer-intensive method to gain insight into the behavior of developed methods or to compare the performance of different methods in a controlled setting. The idea behind Monte Carlo simulation studies is to perform identical computations on a sufficiently large number of independent samples from the same population. Two main categories of simulation studies are thereby distinguished in the literature: *model-based* and *design-based* simulation.

In model-based simulation, samples are generated repeatedly from an infinite population given by a certain distributional model. In every iteration, certain methods are applied and quantities of interest are computed for comparison. Reference values can be obtained from the underlying theoretical distribution where appropriate. This type of simulation is typically performed when the methods of interest make certain assumptions about the distribution of the data. An example is the comparison of outlier detection methods, which typically assume a multivariate normal distribution. For an outline of some general issues in the generation of random numbers, see Section 1.4.1.

Design-based simulation is popular in survey statistics, as samples are drawn repeatedly from a finite population. The *close-to-reality* approach thereby aims to be as realistic as possible, e.g., using the sampling designs applied in real-life for specific surveys such as EU-SILC (*European Union Statistics on Income and Living Conditions*). In every iteration, certain estimators such as indicators are computed, or other procedures such as imputation are applied. The obtained values can then be compared to the true population values where appropriate. Nevertheless, since real population data are in general not available to researchers, synthetic populations need to be generated from existing samples (see Chapter 5). An example for design-based simulation is the comparison of different methods for point- and variance estimation of social inclusion indicators, as performed in the AMELI project (see Section 1.1). The general design of such simulation studies is further discussed in Section 1.4.2

The two approaches described above are also frequently combined. In these so-called *mixed* simulation designs, samples are drawn repeatedly from each generated data set.

1.4.1 Random number generation

The main problem in the generation of random numbers is that computers behave completely deterministic. Nevertheless, many strategies for the generation of *pseudorandom* numbers have been developed to date. Uniform pseudorandom number generators aim to be best possible approximations of a uniform distribution. An important example for such random number generators is the *Mersenne Twister* introduced by Matsumoto and Nishimura

(1998). These uniform pseudorandom numbers can then be further transformed in order to obtain pseudorandom numbers from other distributions. Introductions to random number generation and statistical simulation can be found in, e.g., Morgan (1984), Ripley (1987), Johnson (1987) and Jones et al. (2009).

Real random numbers can be generated based on atmospheric noise (see Haahr 2010). For statistics research, however, reproducibility of the results is a necessity. Results based on real random numbers of course cannot be reproduced. Pseudorandom number generators, on the other hand, can be set to a certain state at any time. Identical computations starting from the same state always produce identical results. Therefore, pseudorandom number generators are in fact favorable for statistical computing.

1.4.2 General design of simulation studies in survey statistics

Close-to-reality simulation studies aim to be as realistic as possible, but they should also be practical with respect to the best possible evaluation of statistical methods. Starting from (synthetic) population data, samples are drawn repeatedly using the real-life sampling methods and weighting schemes. In each simulation run, estimates for certain quantities of interest such as indicators or the associated variance estimates are computed from the corresponding sample. The results of all simulation runs are then combined to form a distribution and are compared to the true values. See also Alfons et al. (2009) for a more detailed discussion on the topic.

Concerning outliers and nonresponse, the most realistic perception of the world is that they exist on the population level. Whether a person has an extremely high income or is not willing to respond to certain questions of a survey does not depend on whether that person is actually in the sample. Hence the most realistic simulation design is to apply modifications such as contamination and nonresponse to the population, as proposed by Münnich et al. (2003b). A diagram depicting this process is shown in Figure 1.3. It is important to note that this approach results in an unpredictable amount of outliers or missing values in the samples, which is a clear disadvantage.

If robustness properties of the considered estimators are the main focus of a simulation, or if outlier detection methods are investigated, maximum control over the amount of contaminated observations is necessary for a thorough evaluation. The same principle applies with respect to the amount of missing values if the treatment of incomplete data is the main interest. In order to solve this problem, Alfons et al. (2009) proposed to add contamination and nonresponse to the samples instead of the population. While this approach may not truly reflect the real processes, it may be more practical from a statistician's point of view. Nevertheless, it should be noted that adding contamination and nonresponse to samples comes with increasing computational costs for an increasing number of simulation runs. Figure 1.4 visualizes the general outline of such a simulation design.

1.4 Statistical simulation

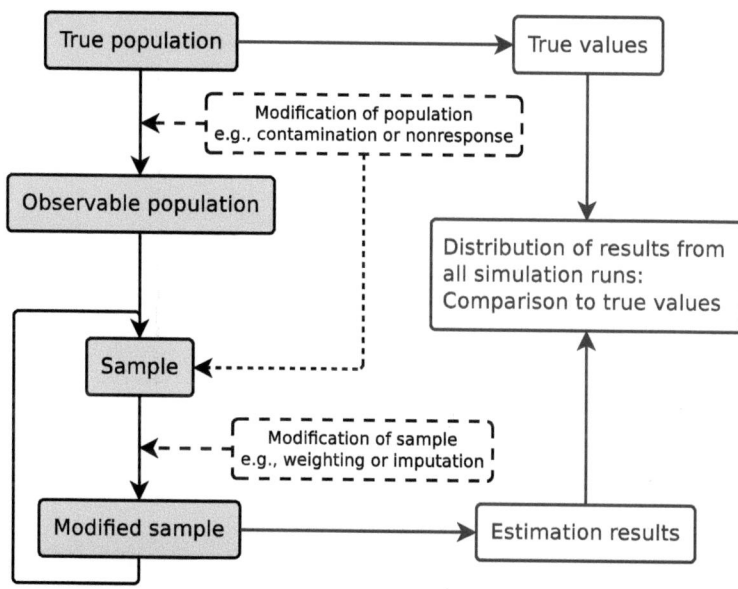

Figure 1.3: Outline of the most realistic design for simulation studies in survey statistics (cf. Münnich et al. 2003b).

Since real population data are in most cases not available to researchers, synthetic population data usually need to be generated to be able to perform close-to-reality simulation studies in survey statistics. These simulated data are generated from an original sample such that the properties of the underlying sample are reflected (see Chapter 5 for details). Hence also nonresponse in the population may be generated with a model from the underlying sample (see Section 1.4.5). However, the situation is much more complicated with contamination. While nonresponse is clearly visible in a sample, the contaminated observations are not known beforehand. Outlier detection methods may be applied, but the results still leave some uncertainty. Even if outliers are detected, it is still necessary to find a model for their distribution, which may or may not depend on the majority of the observations. Simply experimenting with different proportions of outliers and different distributions for the contamination might be a more practical approach. In any case, the design of simulation studies should not be confined to only the most realistic contamination and missing data models. The better the behavior of the developed methodology is known, the better recommendations for its use can be given.

1. Introduction

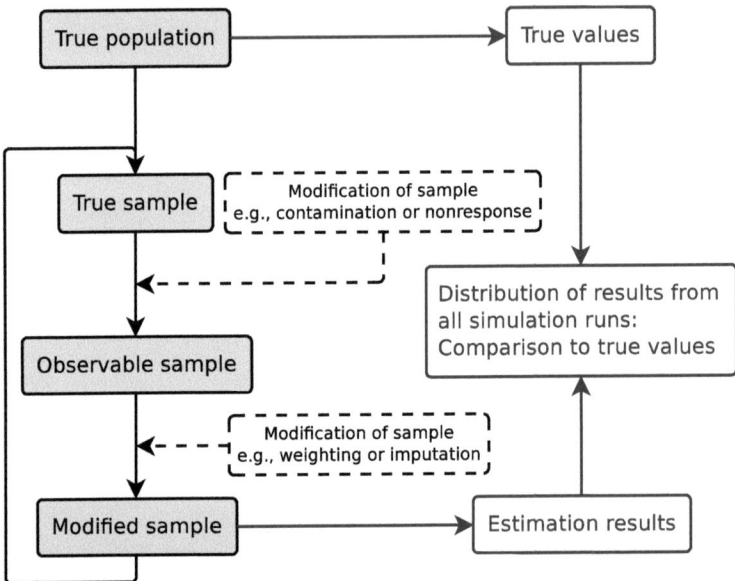

Figure 1.4: Outline of a more practical approach for the simulation studies in survey statistics.

1.4.3 Finite population sampling and weighting

In survey statistics, the observations in the population are considered to contain fixed values that do not vary. Randomness and uncertainty are introduced by the sampling process. As a consequence, many different sampling algorithms from finite populations have been proposed in the literature.

The most basic sampling method is *simple random sampling*, where each element in the population is included in the sample with the same probability. However, often the behavior of different subpopulations is of interest in a survey. In order to ensure that all subpopulations are represented by a sufficient number of observations in the sample, *stratified sampling*, i.e., drawing samples separately from the different subpopulations, may be applied. If the subgroups are relatively homogeneous but there are strong heterogeneities between the subpopulations, stratified sampling reduces the variance of the obtained estimates (see, e.g., Cochran 1977). It is therefore important that variables with strong correlations to the variables of interest are used for stratification. Another important technique frequently applied in practice, mostly for reasons of cost minimization, is *cluster sampling*. Again, the population is divided into different subgroups, but then a sample of the groups is selected and all individuals from the selected groups are included in the sample. Cluster sampling is highly efficient if the inclusion probabilities are approximately proportional to the cluster sizes and there is little variation between the clusters (e.g., Särndal et al. 2003). This idea can be further generalized to *multi-stage* sampling designs, where *secondary sampling units* are then drawn from the *primary sampling units*, etc. For more information on such sampling techniques, the reader is referred to Cochran (1977) and Särndal et al. (2003).

For doing statistical analysis with survey samples, it is often desirable that certain (smaller) population subgroups are overrepresented. Consequently, many different algorithms for taking samples from a finite population with unequal probabilities have been introduced in the literature. Important examples are Midzuno sampling (Midzuno 1952), Poisson sampling (Hájek 1964), Sampford sampling (Sampford 1967), maximum entropy sampling (Chen et al. 1994) and Tillé sampling (Tillé 1996, Deville and Tillé 1998). An overview of a large collection of sampling algorithms is given in Tillé (2006).

When samples are taken from a finite population, each observation in the sample represents a number of observations in the population, as indicated by the *sample weights*. The initial *design weights* are thereby given by $\frac{1}{\pi_i}$, $i = 1, \ldots, n$, where π_i denotes the inclusion probabilities and n is the number of sampled observations. These initial weights are often modified by *calibration* (e.g., Deville et al. 1993) to obtain a set of weights that for certain subgroups sum up to known marginal population totals. In the case of simple random sampling, calibration is performed after *poststratification* (e.g., Cochran 1977). A different approach is *balanced sampling*, which already in the sampling process ensures that certain marginal totals are respected. Deville and Tillé (2004) and Chauvet and Tillé (2006) fairly recently developed a fast algorithm for balanced sampling called the *cube method*.

In close-to-reality simulation studies, the sampling designs and weighting schemes used in real life should certainly be considered. Nevertheless, also other methods should be investigated to see if improvements are possible, e.g., for variance estimation.

1.4.4 Contamination models

When evaluating robust statistical methods in simulation studies, a certain part of the data needs to be contaminated, so that the influence of these outliers on the robust estimators (and possibly their classical counterparts) can be studied. In robust statistics, the distribution of contaminated data is typically modeled as a mixture of distributions as given in (1.9). Consequently, outliers may be modeled by a two-step process in simulation studies (Béguin and Hulliger 2008, Hulliger and Schoch 2009b):

1. Select the observations to be contaminated. If the probabilities of selection do not depend on any information in the data set, the outliers may be called *outlying completely at random* (OCAR), otherwise they may be called *outlying at random* (OAR).

2. Model the distribution of the outliers. If the distribution does not depend on the original values of the selected observations, the contamination may be called *distributed completely at random* (DCAR) or *contaminated completely at random* (CCAR), otherwise it may be called *distributed at random* (DAR) or *contaminated at random* (CAR).

A more detailed mathematical notation of this process is given in Chapter 4, therefore it is omitted here.

1.4.5 Missing data models

This section discusses the use of missing data models in Monte Carlo simulation studies. Missing values are included in many data sets, in particular survey data hardly ever contain complete information. In practice, missing values often need to be imputed, which results in additional uncertainty in further statistical analysis (e.g., Little and Rubin 2002). This additional variability needs to be considered when computing variance estimates or confidence intervals. In simulation studies, it may therefore be of interest to study the properties of different imputation methods or to investigate the influence of missing values on point and variance estimates.

First, the theoretical concepts of different missing data mechanisms are presented. Afterwards, a brief outline on how to incorporate such missing data mechanisms in simulations is given.

Missing data mechanisms

In the missing data literature, three important cases of processes generating missing values are commonly distinguished, based on ideas by Rubin (1976). For a more detailed discussion on these missing data mechanisms, the reader is referred to Little and Rubin (2002).

Let $\boldsymbol{X} = (x_{ij})_{1 \leq i \leq n, 1 \leq j \leq p}$ denote the data, where n is the number of observations and p the number of variables, and let $\boldsymbol{M} = (M_{ij})_{1 \leq i \leq n, 1 \leq j \leq p}$ be an indicator whether an observation is missing ($M_{ij} = 1$) or not ($M_{ij} = 0$). Furthermore, let the observed and missing parts of the data be denoted by $\boldsymbol{X}_{obs}$ and $\boldsymbol{X}_{miss}$, respectively. The missing data mechanism is then characterized by the conditional distribution of $\boldsymbol{M}$ given $\boldsymbol{X}$, denoted by $f(\boldsymbol{M}|\boldsymbol{X}, \boldsymbol{\phi})$, where $\boldsymbol{\phi}$ denotes unknown parameters (see Little and Rubin 2002).

The missing values are *missing completely at random* (MCAR) if the missingness does not depend on the data $\boldsymbol{X}$, i.e., if

$$f(\boldsymbol{M}|\boldsymbol{X}, \boldsymbol{\phi}) = f(\boldsymbol{M}|\boldsymbol{\phi}). \tag{1.12}$$

Note that there may still be a certain pattern in the missing values, depending on the unknown parameters $\boldsymbol{\phi}$, but such a pattern will be independent of the actual data. A more

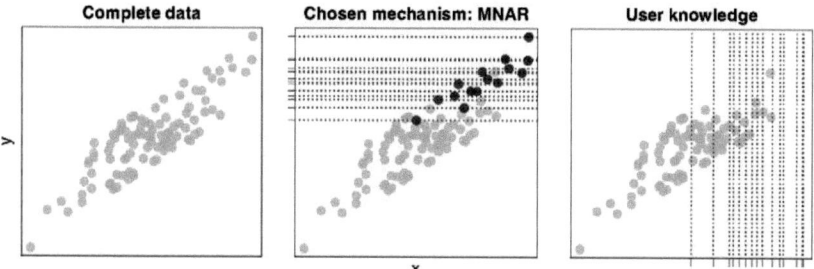

Figure 1.5: Simulated bivariate data set with high correlation. *Left*: Complete data. *Center*: Black points are chosen as missing in y, depending on the value of y (MNAR). *Right*: Information is only available for x-values in practice.

general scenario is given if the missingness depends on the observed information X_{obs}. In this case, the missing values are *missing at random* (MAR), which translates to the equation

$$f(M|X,\phi) = f(M|X_{obs},\phi). \tag{1.13}$$

On the other hand, the missing values are said to be *missing not at random* (MNAR) if Equation (1.13) is violated. This can be written as

$$f(M|X,\phi) = f(M|(X_{obs},X_{miss}),\phi). \tag{1.14}$$

Hence, in the latter case, the missing values cannot be fully explained by the observed part of the data.

A motivational example for the different missing data mechanisms is given in Little and Rubin (2002). Consider two variables *age* and *income*, with missing values in income. If the probability of missingness is the same for all individuals, regardless of their age or income, then the data are MCAR. If the probability that income is missing varies according to the age of the respondent, but does not vary according to the income of respondents with the same age, then the data are MAR. If the probability that income is recorded varies according to income for those with the same age, then the data are MNAR. Note that MNAR *cannot* be detected in practice, as this would require knowledge of the missing values themselves. This problem is further illustrated in Figures 1.5 and 1.6 using simulated bivariate data sets.

Figure 1.5 shows a highly correlated bivariate data set with variables x and y. From the complete data (*left*), the y-part of some observations are marked as missing depending on the value of y (*center*). The probability of missingness is higher for larger values in y, hence the missing values in this example are constructed as MNAR. Nevertheless, in practice only

1. Introduction

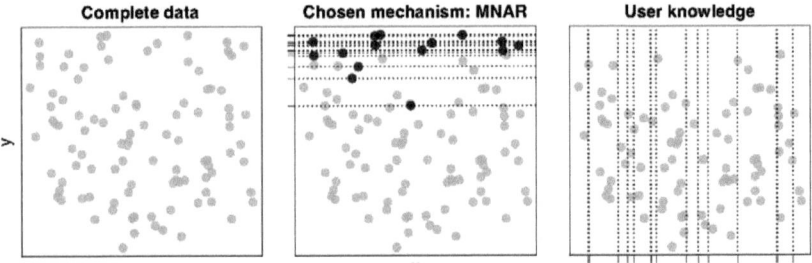

Figure 1.6: Simulated bivariate data set without correlation. *Left*: Complete data. *Center*: Black points are chosen as missing in y, depending on the value of y (MNAR). *Right*: Information is only available for x-values in practice.

the x-part of the observations with missing values in y are known (*right*), i.e., it is only observable that the amount of missing values increases for increasing x-values. Therefore, a MAR situation would be assumed knowing that this could also be an MNAR situation. In other words, it is impossible to distinguish between MAR and MNAR due to the high correlation between x and y.

A similar picture is displayed in Figure 1.6, but with uncorrelated variables. Again, the y-part of some observations is marked as missing depending on the value of y (MNAR). In practice, however, the probability of missingness seems to be completely independent from the data values, therefore an MCAR situation would be detected. Since an MNAR situation cannot be confirmed or ruled out, it is impossible to distinguish between MCAR and MNAR in the case of uncorrelated variables.

The performance of imputation methods usually depends, among other things, on the multivariate structure of the missing values. In simulation studies in survey statistics, realistic missing data scenarios should be investigated. Therefore, the existing real-life samples of the survey of interest need to be studied. The R package **VIM** (Templ and Filzmoser 2008, Templ et al. 2010a) contains visualization tools for exploring incomplete data that allow not only to detect the missing value mechanisms (MAR or MCAR), but also to gain insight into the quality and various other aspects of the data at the same time.

Adding missing data in simulations

Choosing variables in which missing values should be inserted is the first step of adding missing data in a simulation study. These variables will in the following be referred to as *target* variables.

Random insertion of missing values: MCAR situations can be generated by selecting observations for every target variable with simple random sampling. Using unequal probability sampling makes it possible to account for MAR or MNAR situations. The inclusion probabilities may thereby be derived from one or more variables. In any case, different missing value rates may be used for the different target variables. More difficult situations require special treatment, though. An important example is that missing values in one variable may only occur for observations with missing values in another variable.

Response propensity models: In simulation studies in survey statistics, a completely different approach for generating MAR situations is to use a logit model, which can be obtained from the simulation's underlying real-life sample. With the imputed original sample, even MNAR situations may be generated. More details on response propensity models with several applications can be found in Münnich et al. (2003a). The advantage of this method is that it is a more analytical approach for generating realistic nonresponse scenarios. However, the resulting number of missing values is not predictable. Thus response propensity models are more suitable for adding nonresponse to population data than to samples, as having maximum control over the number of missing values is the motivation for the latter.

1.4.6 Parallel computing

Monte Carlo simulation is based on carrying out identical computations on a sufficiently large number of independent samples from the same population (either an infinite population given by a certain theoretical distribution or a finite population). Hence such simulations are an *embarrassingly parallel* process, i.e., computational performance can be increased by parallel computing. As nowadays nearly every computer in use has multiple processor cores or even CPUs, parallel computing is an ever growing field of research. Nevertheless, there are some issues regarding the random number generator that need to be considered when using parallel computing for simulation studies in statistics.

If computations are spread out over a number of parallel processes and standard sequential random number generators are used, there may be overlaps in the produced sequences of random numbers. In the worst case, the sequential random number generators are initialized with the same seed and identical sequences are generated in each of the parallel processes, which for Monte Carlo simulation means that identical results are produced. Combining the simulation results from the different processes in order to form a distribution then clearly suffers from a loss of efficiency compared to producing the same number of results in a single process. This loss of efficiency is due to the fact that the samples are no longer independent of each other. However, simply initializing the sequential random number generators with a different or random seed is not enough to avoid this problem. One generator may at some point reach a certain state that another generator was in at a previous step. Then the remaining sequence of random numbers produced by the first generator has also been produced by the second one, thus the requirement of independence is again violated.

As a remedy, parallel random number generators that produce multiple independent streams of random numbers have been developed. Important examples are the generators developed by L'Ecuyer et al. (2002) and Mascagni and Srinivasan (2000). When performing statistical simulation by parallel computing, it is thus absolutely essential that such parallel random number streams are used.

1.5 Outline of the remaining chapters

Chapters 2–4 introduce a general framework for statistical simulation that has been implemented in the statistical environment R (R Development Core Team 2010) as the add-on package **simFrame** (Alfons 2010). In particular, Chapter 2 discusses the implementation in great detail and has been accepted for publication in the *Journal of Statistical Software* (Alfons et al. 2010c). Chapter 3 is supplementary material to the paper from Chapter 2 and will be available on the website of the *Journal of Statistical Software*. It contains additional code examples demonstrating the strengths and usefulness of the developed framework. In Chapter 4, which has been published in the conference proceedings *Computer Data Analysis and Modeling: Complex Stochastic Data and Systems* (Alfons et al. 2010d), the contamination models implemented in **simFrame** for the evaluation of robust statistical methods are described in more detail.

For close-to-reality simulation studies in survey statistics, synthetic population data are necessary. Therefore, a method for the simulation of population data for complex household surveys such as EU-SILC has been developed and implemented in the R package **simPopulation** (Alfons and Kraft 2010). The statistical methodology is described in detail in Chapter 5, which has been submitted to the journal *Statistical Methods & Applications*. An earlier version of the paper is available as technical report (Alfons et al. 2010b). Data confidentiality issues of such synthetic population data are then discussed in Chapter 6, which has been published in *Privacy in Statistical Databases*, volume 6344 of *Lecture Notes in Computer Science* (Templ and Alfons 2010).

Chapter 7 applies the simulation methodology from the previous chapters to investigate the use of robust Pareto tail modeling to reduce the influence of outliers on selected Laeken indicators. It is a slightly corrected version of a paper published in *Combining Soft Computing and Statistical Methods in Data Analysis*, volume 77 of *Advances in Intelligent and Soft Computing* (Alfons et al. 2010e). The selected Laeken indicators and methods for Pareto tail modeling have been implemented in the R package **laeken** (Alfons et al. 2010a).

Another application of robust statistics in a related topic is presented in Chapter 8, which has been accepted for publication in the journal *Statistical Methods & Applications*

(Alfons et al. 2011a). More precisely, a robust variable selection procedure for applications in the social sciences has been developed and demonstrated in an application to quality of life research. The R code for the procedure is freely available from http://www.statistik.tuwien.ac.at/public/filz/programs.html. It could not be published as a package, as it requires code by Khan et al. (2007b) that is only available from the website of these authors as well.

1.5.1 Overview of the remaining chapters

To summarize the contents of this thesis, this section provides a short overview of the remaining chapters. In addition, it is indicated whether or where they have already been published.

Chapter 2
> A. Alfons, M. Templ, and P. Filzmoser. An object-oriented framework for statistical simulation: The R package **simFrame**. *Journal of Statistical Software*. Accepted for publication.

Chapter 3
> A. Alfons, M. Templ, and P. Filzmoser. Applications of statistical simulation in the case of EU-SILC: Using the R package **simFrame**. *Journal of Statistical Software*. Supplementary material to the paper from Chapter 2.

Chapter 4
> A. Alfons, M. Templ, and P. Filzmoser. Contamination models in the R package **simFrame** for statistical simulation. In S. Aivazian, P. Filzmoser, and Y. Kharin (editors), *Computer Data Analysis and Modeling: Complex Stochastic Data and Systems*, volume 2, pages 178–181, Minsk, 2010. ISBN 978-985-476-848-9.

Chapter 5
> A. Alfons, S. Kraft, M. Templ, and P. Filzmoser. Simulation of close-to-reality population data for household surveys with application to EU-SILC. Revision submitted to *Statistical Methods & Applications*.

Chapter 6
> M. Templ and A. Alfons. Disclosure risk of synthetic population data with application in the case of EU-SILC. In J. Domingo-Ferrer and E. Magkos (editors), *Privacy in Statistical Databases*, volume 6344 of *Lecture Notes in Computer Science*, pages 174–186. Springer, Heidelberg, 2010. ISBN 978-3-642-15837-7.

1. Introduction

Chapter 7
A. Alfons, M. Templ, P. Filzmoser, and J. Holzer. A comparison of robust methods for Pareto tail modeling in the case of Laeken indicators. In C. Borgelt, G. González-Rodríguez, W. Trutschnig, M.A. Lubiano, M.A. Gil, P. Grzegorzewski, and O. Hryniewicz (editors), *Combining Soft Computing and Statistical Methods in Data Analysis*, volume 77 of *Advances in Intelligent and Soft Computing*, pages 17–24. Springer, Heidelberg, 2010. ISBN 978-3-642-14745-6.

Chapter 8
A. Alfons, W.E. Baaske, P. Filzmoser, W. Mader, and R. Wieser. Robust variable selection with application to quality of life research. *Statistical Methods & Applications*, pages 1–18, 2010. http://dx.doi.org/10.1007/s10260-010-0151-y. DOI 10.1007/s10260-010-0151-y, to appear.

Chapter 2

An object-oriented framework for statistical simulation: The R package simFrame

Accepted for publication in the *Journal of Statistical Software* (Alfons et al. 2010c).

Andreas Alfons[a], Matthias Templ[a,b], Peter Filzmoser[a]

[a] Department of Statistics and Probability Theory, Vienna University of Technology
[b] Methods Unit, Statistics Austria

Abstract Simulation studies are widely used by statisticians to gain insight into the quality of developed methods. Usually some guidelines regarding, e.g., simulation designs, contamination, missing data models or evaluation criteria are necessary in order to draw meaningful conclusions. The R package **simFrame** is an object-oriented framework for statistical simulation, which allows researchers to make use of a wide range of simulation designs with a minimal effort of programming. Its object-oriented implementation provides clear interfaces for extensions by the user. Since statistical simulation is an embarrassingly parallel process, the framework supports parallel computing to increase computational performance. Furthermore, an appropriate plot method is selected automatically depending on the structure of the simulation results. In this paper, the implementation of **simFrame** is discussed in great detail and the functionality of the framework is demonstrated in examples for different simulation designs.

Keywords R, statistical simulation, outliers, missing values, parallel computing

2.1 Introduction

Due to the complexity of modern statistical methods, obtaining analytical results about their properties is often virtually impossible. Therefore, simulation studies are widely used by statisticians as data-based, computer-intensive alternatives for gaining insight into the quality of developed methods. However, research projects commonly involve many scientists, often from different institutions, each focusing on different aspects of the project. If these researchers use different simulation designs, the results may be incomparable, which in turn makes it impossible to draw meaningful conclusions. Hence simulation studies in such research projects require a precise outline.

The R package **simFrame** (Alfons 2010) is an object-oriented framework for statistical simulation addressing this problem. Its implementation follows an object-oriented approach based on S4 classes and methods (Chambers 1998, 2008). A key feature is that statisticians can make use of a wide range of simulation designs with a minimal effort of programming. The object-oriented implementation gives maximum control over input and output, while at the same time providing clear interfaces for extensions by user-defined classes and methods.

Comprehensive literature exists on statistical simulation, but is mainly focused on technical aspects (e.g., Morgan 1984, Ripley 1987, Johnson 1987). Unfortunately, hardly any publications are available regarding the conceptual elements and general design of modern simulation experiments. To name some examples, Münnich et al. (2003b) and Alfons et al. (2009) describe how close-to-reality simulations may be performed in survey statistics, while Burton et al. (2006) address applications in medical statistics. Furthermore, while simulation studies are widely used in scientific articles, they are often described only briefly and without sufficient details on all the processes involved. Having a framework with different simulation designs ready at hand may help statisticians to plan simulation studies for their needs.

Statistical simulation is frequently divided into two categories: *design-based* and *model-based* simulation. Design-based simulation is popular in survey statistics, as samples are drawn repeatedly from a finite population. The close-to-reality approach thereby uses the true sampling designs for real-life surveys such as EU-SILC (European Union Statistics on Income and Living Conditions). In every iteration, certain estimators such as indicators are computed or other statistical procedures such as imputation are applied. The obtained values can then be compared to the true population values where appropriate. Nevertheless, since real population data is only in few cases available to researchers, synthetic populations may be generated from existing samples (see, e.g., Münnich et al. 2003b, Münnich and Schürle 2003, Raghunathan et al. 2003, Alfons et al. 2010b). Such synthetic populations must reflect the structure of the underlying sample regarding dependencies among the variables and heterogeneity. For household surveys, population data can be generated using the R package **simPopulation** (Alfons and Kraft 2010). In model-based simulation, on the other hand, data sets are generated repeatedly from a distributional model or a mixture of distributions. In every iteration, certain methods are applied and quantities of interest are computed for comparison. Where appropriate, reference values can be obtained from the underlying theoretical distribution. *Mixed* simulation designs constitute a combination of the two approaches, in which samples are drawn repeatedly from each generated data set.

The package **simFrame** is intended to be as general as possible, but has initially been developed for close-to-reality simulation studies in survey statistics. Moreover, it is focused on simulations involving typical data problems such as outliers and missing values. Therefore, certain proportions of the data may be contaminated or set as missing in order to investigate the quality and behavior of, e.g., robust estimators or imputation methods. In addition, an appropriate plot method for the simulation results is selected automatically depending on their structure. Note that statistical simulation is a very loose concept, though, and that the application of **simFrame** may be subject to limitations in certain scenarios.

Section 2.2 gives a brief introduction to the basic concepts of object-oriented programming and the S4 system. In Section 2.3, the design of the framework is motivated and Section 2.4 describes the implementation in great detail. Section 2.5 then provides details about parallel computing with **simFrame**. The use of the package for different simulation designs is demonstrated in Section 2.6. Additional examples for design-based simulation are given in a supplementary paper. How to extend the framework is outlined in Section 2.7. Finally, Section 2.8 contains concluding remarks and gives an outlook on future developments.

2.2 Object-oriented programming and S4

The object-oriented paradigm states that problems are formulated using interacting objects rather than a set of functions. The properties of these objects are defined by *classes* and their behavior and interactions are modeled with *generic functions* and *methods*. One of the most important concepts of object-oriented programming is *class inheritance*, i.e., *subclasses* inherit properties and behavior from their *superclasses*. Thus code can be shared for related classes, which is the main advantage of inheritance. In addition, subclasses may have additional properties and behavior, so in this sense they *extend* their superclasses. In S4 (Chambers 1998, 2008), properties of objects are stored in *slots* and can be accessed or modified with the @ operator or the slot() function. However, *accessor* and *mutator* methods are supposed to be used to access or modify properties of objects in **simFrame** (see Section 2.3.3). *Virtual classes* are special classes from which no objects can be created. They exist for the sole reason of sharing code. Furthermore, *class unions* are special virtual classes with no slots.

2. An object-oriented framework for statistical simulation

Generic functions define the formal arguments that are evaluated in a function call in order to select the actual method to be used. These methods are defined by their *signatures*, which assign classes to the formal arguments. In short, generic functions define *what* should be done and methods define *how* this should be done for different (combinations of) classes. As an example, the generic function `setNA()` is used in **simFrame** to insert missing values into a data set. These are the available methods:

```
R> showMethods("setNA")

Function: setNA (package simFrame)
x="data.frame", control="character"
x="data.frame", control="missing"
x="data.frame", control="NAControl"
```

Even though a simple object-oriented mechanism was introduced in S3 (Chambers and Hastie 1992), it is not sufficient for the purpose of implementing a flexible framework for statistical simulation. Only S4 offers consequent implementations of advanced object-oriented techniques such as inheritance, object validation and method signatures. In S3, inheritance is realized by simply using a vector for the `class` attribute, hence there is no way to guarantee that the subclass contains all properties of the superclass. It should be noted that the tradeoff of these advanced programming techniques is a slightly increased computational overhead. Nevertheless, with modern computing power, this is not a substantial issue.

2.3 Design of the framework

Statistical simulation in R (R Development Core Team 2010) is often done using bespoke use-once-and-throw-away scripts, which is perfectly fine when only a handful of simulation studies need to be done for a specific purpose such as a paper. But when a research project is based on extensive simulation studies with many different simulation designs, this approach has its limitations since substantial changes may need to be applied to the R scripts for each design. In addition, if many partners are involved in the project and each of them writes their own scripts, they need to be very well coordinated so that the implemented simulation designs are similar, otherwise the obtained results may not be comparable.

The fundamental design principle of **simFrame** is that the behavior of functions is determined by *control objects*. A collection of such control objects, including a function to be applied in each iteration, is simply plugged into a generic function called `runSimulation()`, which then performs the simulation experiment. This allows to easily switch from one simulation design to another by just plugging in different control objects. Note that the user does not have to program any loops for iterations or collect the results in a suitable data structure, the framework takes care of this. Furthermore, by using the package as a common framework for simulation in research projects, guidelines for simulation studies may be defined by selecting specific control classes and parameter settings. If the researchers decide on a set of control objects to be used in the simulation studies, this ensures comparability of the obtained results and avoids problems with drawing conclusions from the project. Defining control objects thereby requires only a few lines of code, and storing them as RData files in order to distribute them among partners is much easier than ensuring that a large number of R scripts with big chunks of bespoke code are comparable.

As a motivational example, consider a research project in which researchers A and B investigate a specific survey such as EU-SILC (European Union Statistics on Income and Living Conditions). Researcher A focuses on robust estimation of certain indicators, while researcher B tries to improve the data quality with more suitable imputation and outlier detection methods. The aim of the project is to evaluate the developed methods with extensive simulation studies. In order to be as realistic as possible, design-based simulation studies are performed, where samples are drawn repeatedly from (synthetic) population data. Let the survey of interest in real life be conducted in many countries with different sampling designs. Then A and B could each define some control objects for the most common sampling designs and exchange them so that they can plug each of them into the function runSimulation() along with the population data.

Since imputation methods and outlier detection methods typically make some theoretical assumptions about the data, B could also carry out model-based simulation studies, in which the data are repeatedly generated from a certain theoretical distribution. All B needs to change is to define a control object to generate the data and supply it to runSimulation() instead of the population data and the control object for sampling.

Both researchers in this example investigate robust methods. It may be of interest to explore the behavior of these methods under different contamination models (the term *contamination* is used in a technical sense in this paper, see Section 2.4.3 for a definition). This can again be done by defining and exchanging a set of control objects. In addition, B can define various control objects for inserting missing values into the data in order to study the performance of imputation methods. Switching from one contamination model or missing data mechanism to another is simply done by replacing the respective control object in the call to runSimulation(). B could also supply a control object for inserting contamination and one for inserting missing values to investigate robust imputation methods or outlier detection methods for incomplete data.

One example for such research projects is the project AMELI (Advanced Methodology of European Laeken Indicators, http://ameli.surveystatistics.net), in the course of which the package **simFrame** has been developed.

2.3.1 UML class diagram

The Unified Modeling Language (UML) (Fowler 2003) is a standardized modeling language used in software engineering. It provides a set of graphical tools to model object-oriented programs. A *class diagram* visualizes the structure of a software system by showing classes, attributes, and relationships between the classes. Figure 2.1 shows a slightly simplified UML class diagram of **simFrame**.

In this example, classes are represented by boxes with two parts. The top part contains the name of the class and the bottom part lists its slots. Class names in italics thereby indicate virtual classes. Furthermore, each slot is followed by the name of its class, which can be a basic R data type such as `numeric`, `character`, `logical`, `list` or `function`, but also an S4 class.

Lines or arrows of different forms represent class relationships. Inheritance is denoted by an arrow with an empty triangular head pointing to the superclass. Composition, i.e., a class having another class as a slot, is depicted by an arrow with a solid black diamond on the side of the composed class. A solid line indicates an *association* between two classes. Here an association signals that there is a method with one class as primary input and the other class as output. Last but not least, a dashed line denotes an *association class*, which in the case of **simFrame** is a control class that is not the primary input of the corresponding method but nevertheless determines its behavior.

2.3.2 Naming conventions

In order to facilitate the usage of the framework, the following naming rules are introduced:

- Names of classes, slots, functions and methods are alphanumeric in mixed case, where the first letter of each internal word is capitalized.

- Class names start with an uppercase letter.

- Functions, methods and slots start with a lowercase letter. Exceptions are functions that initialize a class, which are called *constructors* and have the same name as the class.

- Violate the above rules whenever necessary to maintain compatibility.

These rules are based on code conventions for the programming language Java (e.g., Arnold et al. 2005), see `http://java.sun.com/docs/codeconv/`. Some R packages, e.g., **rrcov** (Todorov and Filzmoser 2009, Todorov 2010), use similar rules.

2.3 Design of the framework

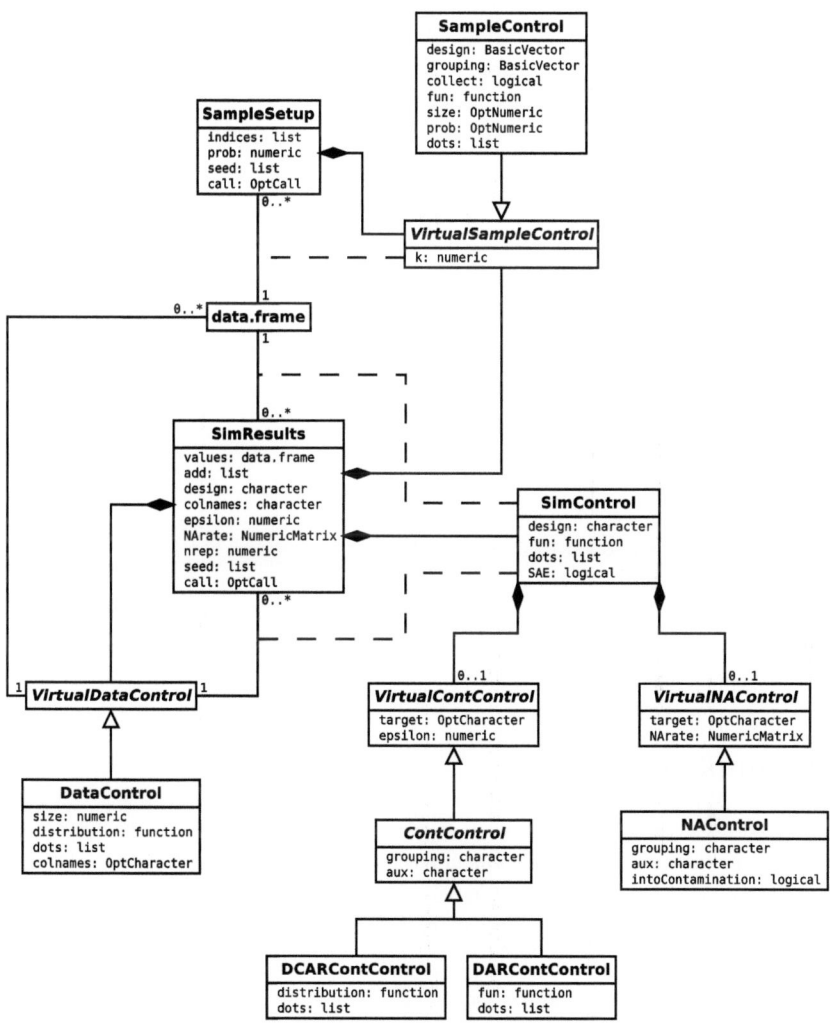

Figure 2.1: Slightly simplified UML class diagram of **simFrame**.

2.3.3 Accessor and mutator methods

In object-oriented programming languages, *accessor* and *mutator* methods are typically used to retrieve and change the properties of a class, respectively. The idea behind this concept is to hide information about the actual implementation of a class (e.g., what data structures are used) from the user. In **simFrame**, accessors are named `getFoo()` and mutators are named `setFoo()`, where `foo` is the name of the corresponding slot. This naming convention is common in Java and is also used in some R packages (e.g., **rrcov**).

The use of accessor and mutator methods in **simFrame** is illustrated with the class `NAControl`, which handles the insertion of missing values into a data set (see Section 2.4.4). Its slot `NArate` controls the proportion of missing values to be inserted.

```
R> nc <- NAControl(NArate = 0.05)
R> getNArate(nc)
```

```
[1] 0.05
```

```
R> setNArate(nc, c(0.01, 0.03, 0.05, 0.07, 0.09))
R> getNArate(nc)
```

```
[1] 0.01 0.03 0.05 0.07 0.09
```

Note that if no method `setFoo()` is available, the slot is not supposed to be changed by the user. However, as already mentioned in Section 2.2, R allows every slot to be modified with the @ operator or the `slot()` function.

2.4 Implementation

The open-source statistical environment R has become the main framework for computing in statistics research. One of its main advantages is that it includes a well-developed programming language and provides interfaces to many others, including the fast low-level languages C and Fortran. The S4 system (Chambers 1998, 2008) complies with all requirements for an object-oriented framework for statistical simulation. Thus most of **simFrame** is implemented as S4 classes and methods, except some utility functions and some C code.

Method selection for generic functions is based on *control classes*, which in most cases provides the interfaces for extensions by developers (see Section 2.7). Most of these generic functions are not expected to be called by the user directly. The idea of the framework is rather to define a number of control objects and to supply them to the function `runSimulation()`, which performs the whole simulation experiment and calls the other functions internally (see Section 2.4.5 or the examples in Section 2.6).

2.4.1 Data handling

In R, data are typically stored in a data.frame, and **simFrame** is no exception. However, when samples are taken from a finite population in design-based simulation studies, each observation in the sample represents a number of observations in the population, given by the sample weights. Unless a basic sampling procedure such as simple random sampling is used, the weights are in general not equal for all sampled observations and need to be considered to obtain unbiased estimates. But even if the weights are equal for all observations, they may be needed for the estimation of population totals (e.g., the total turnover of all businesses in a country). In practice, the initial weights are also frequently modified by calibration (e.g., Deville et al. 1993), which for simple random sampling is done after post-stratification (e.g., Cochran 1977). Therefore, the sample weights need to be stored.

In addition, the package has been designed with special emphasis on simulations involving typical data problems such as outliers and missing values. It offers mechanisms to contaminate the data and insert missing values so that the influence of these data problems on statistical methods can be investigated, or that outlier detection or imputation methods can be evaluated. The term *contamination* is used in a technical sense here (see Section 2.4.3 for a definition). Information on which observations are contaminated is often required, both for the user running simulations and for internal use. Since it cannot be retrieved from the data otherwise, it needs to be saved.

As a result, additional variables are added to the data set in these situations. The names of the additional variables are ".weight" and ".contaminated", respectively. Hence these column names should be avoided (which is why they start with a dot), or else the corresponding columns will be overwritten.

Statistical methods often make assumptions about the distribution of the data, e.g., outlier detection methods in multivariate statistics usually assume that the majority of the data follow a multivariate normal distribution. Consequently, such methods are typically tested in simulations on data coming from a certain theoretical distribution. The generation of data from a distributional model is handled by control classes inheriting from the class union (which is a special virtual class with no slots) VirtualDataControl. This virtual class is available so that the framework can be extended by the user (see Section 2.7.1). A simple control class already implemented in **simFrame** is DataControl. It consists of the following slots (see also Figure 2.1):

size: The number of observations to be generated.

distribution: A function for generating the data, e.g., rmvnorm in package **mvtnorm** (Genz and Bretz 2009, Genz et al. 2010) for data following a multivariate normal distribution. It should take a positive integer as its first argument (the slot size will be passed) and return an object that can be coerced to a data.frame.

2. An object-oriented framework for statistical simulation

dots: Additional arguments to be passed to `distribution`.

The following example demonstrates how to define a control object for generating data from a multivariate normal distribution.

```
R> library("mvtnorm")
R> dc <- DataControl(size = 10, distribution = rmvnorm, dots =
+      list(mean = rep(0, 2), sigma = matrix(c(1, 0.5, 0.5, 1), 2, 2)))
```

In a model-based simulation study, such a control object is then used by the framework in repeated internal calls of the generic function `generate(control, ...)`.

```
R> foo <- generate(dc)
R> foo
            V1          V2
1  -1.289442540 -0.7733444
2   0.009574644 -0.8925633
3   0.846580448 -0.4691297
4  -2.249086822 -0.5448488
5   0.662837956  1.0378657
6   0.460268469  0.0244493
7  -0.687415163 -0.6423509
8  -0.763840555 -1.0216259
9  -0.761894738 -0.9547367
10 -0.234446371  2.1031588
```

While the function `generate()` is designed to be called internally by the simulation framework, it is possible to use it as a general wrapper function for data generation in other contexts. For convenience, the name of the control class may then also be passed to `generate()` as a character string (the default is `"DataControl"`), in which case the slots may be supplied as arguments. Nevertheless, it might be simpler for the user to call the underlying function from the slot `distribution` directly in such applications.

Memory-efficient storage of data frames has recently been added to package **ff** (Adler et al. 2010), which might be useful for design-based simulation with large population data. The incorporation into **simFrame** may therefore be investigated as a future task.

2.4.2 Sampling

A fundamental design principle of **simFrame** in the case of design-based simulation studies is that the sampling procedure is separated from the simulation procedure. Two main advantages arise from *setting up* all samples in advance.

2.4 Implementation

First, the repeated sampling reduces overall computation time dramatically in certain situations, since computer-intensive tasks like stratification need to be performed only once. This is particularly relevant for large population data. As an example, consider the AMELI project that has been mentioned in Section 2.3. In the close-to-reality simulation studies carried out in this project, up to 10 000 samples are drawn from a population of more than 8 000 000 individuals with stratified sampling or even more complex sampling designs. For such large data sets, stratification takes a considerable amount of time and is a very memory-intensive task. If the samples are taken on-the-fly, i.e., in every simulation run one sample is drawn, the function to take the stratified sample would typically split the population into the different strata in each of the 10 000 iterations. If all samples are drawn in advance, on the other hand, the population data need to be split only once and all 10 000 samples can be taken from the respective strata together.

Second, the samples can be stored permanently, which simplifies the reproduction of simulation results and may help to maximize comparability of results obtained by different partners in a research project. Consider again the AMELI project, where one group of researchers investigates robust semiparametric approaches to improve the estimation of certain indicators (i.e., a distribution is fitted to parts of the data; see Alfons et al. 2010e), while another group is focused on nonparametric methods (e.g., trimming or M-estimators; see Hulliger and Schoch 2009a). The aim of this project is to evaluate these methods in realistic settings, therefore the most commonly used sampling designs in real life are applied in the simulation studies. If the two groups use not only the same population data, but also the same previously set up samples, their results are highly comparable. In addition, the same samples may be used for other close-to-reality simulation studies within the project, e.g., in order to evaluate imputation or outlier detection methods. This is useful in particular for large population data, when complex sampling techniques may be very time-consuming.

In **simFrame**, the generic function `setup(x, control, ...)` is available to set up multiple samples. It returns an object of class `SampleSetup`, which contains the following slots (among others, all slots are shown in Figure 2.1):

indices: A list containing the indices of the sampled observations.

prob: A numeric vector giving the inclusion probabilities for every observation of the population. These are necessary to compute the sample weights.

seed: A list containing the seeds of the random number generator before and after setting up the samples, respectively.

The function `setup()` may be called by the user to permanently store the samples, but it may also be called internally by the framework if this is not necessary. In any case, methods are selected according to control classes extending `VirtualSampleControl`, which

2. An object-oriented framework for statistical simulation

is a virtual class whose only slot k specifies the number of samples to be set up. This virtual class provides the interface for extensions by the user (see Section 2.7.2). The implemented control class `SampleControl` is highly flexible and covers the most frequently used sampling designs in survey statistics:

- Sampling of individual observations with a basic sampling method such as simple random sampling or unequal probability sampling.

- Sampling of whole groups (e.g., households) with a specified sampling method. There are two common approaches towards sampling of groups:

  – Groups are sampled directly. This is usually referred to as *cluster sampling*. However, here the term *cluster* is avoided in the context of sampling to prevent confusion with computer clusters for parallel computing (see Section 2.5 and the example in Section 2.6.3).

  – In a first step, individuals are sampled. Then all individuals that belong to the same group as any of the sampled individuals are collected and added to the sample.

- Stratified sampling using one of the above procedures in each stratum.

In addition to the inherited slot k, the class `SampleControl` consists of the following slots (see also Figure 2.1):

design: A vector specifying variables to be used for stratification.

grouping: A character string, single integer or logical vector specifying a variable to be used for grouping.

collect: A logical indicating whether groups should be collected after sampling individuals or sampled directly. The default is to sample groups directly.

fun: A function to be used for sampling (the default is simple random sampling). For stratified sampling, this function is applied to each stratum.

size: The sample size. For stratified sampling, this should be a numeric vector.

prob: A numeric vector giving probability weights.

dots: Additional arguments to be passed to `fun`.

Currently, the functions `srs` and `ups` are implemented in **simFrame** for simple random sampling and unequal probability sampling, respectively, but this can easily be extended with user-defined sampling methods (see Section 2.7.2). Note that the sampling method is evaluated using try(). Hence, if an error occurs in obtaining one sample, the others are not lost. This is particularly useful for complex and time-consuming sampling procedures, as the whole process of setting up all samples does not have to be repeated.

The control class for setup() may be specified as a character string (the default is, of course, "SampleControl"), which allows the slots to be supplied as arguments. Furthermore, simSample() is a convenience wrapper for setup() with control class SampleControl.

To actually draw one of the previously set up samples from the population, the generic function draw(x, setup, ...) is used internally by the framework in the simulation runs. It is important to note that the column ".weight", which contains the sample weights, is added to the resulting data set. When sampling from finite populations, storing the sample weights is essential. In general, the weights are not equal for all sampled observations, depending on the inclusion probabilities. Hence the sample weights need to be considered in order to obtain unbiased estimates. But even for simple random sampling, when all weights are equal, each observation in the sample represents a number of observations in the population. For the estimation of population totals (e.g., the total turnover of all businesses in a country), the sample weights are thus still necessary. Moreover, the initial sample weights are in practice often modified by calibration (e.g., Deville et al. 1993). In the case of simple random sampling, this is done after post-stratification (e.g., Cochran 1977).

In the following illustrative example, two samples from synthetic EU-SILC population data are set up and stored in an object of class SampleSetup. EU-SILC is a well-known survey on income and living conditions conducted in European countries (see Section 2.6.1 for more information and a more detailed example). Afterwards, the first of the two set up samples is drawn from the population.

```
R> data("eusilcP")
R> set <- setup(eusilcP, size = 10, k = 2)
R> summary(set)

2 samples of size 10 are set up

R> set

Indices of observations for each of the 2 samples:

[[1]]
 [1] 32456 37914 18290 36471 19342 29442 39711 28444 14306 44891

[[2]]
 [1]  4328 18165 42070 29593  8974 29556 28970 44056 10243 49754

R> draw(eusilcP[, c("id", "eqIncome")], set, i = 1)
```

```
         id eqIncome .weight
33670 1376802 17760.93  5865.4
10460 1611302 13603.65  5865.4
48757 0782302 13910.35  5865.4
33719 1549903 14641.86  5865.4
55360 0825101 12463.90  5865.4
35123 1251602 27331.19  5865.4
29673 1686001 27981.36  5865.4
26985 1210302  7247.55  5865.4
38182 0611002 19968.32  5865.4
6448  1913501 18091.90  5865.4
```

2.4.3 Contamination

When evaluating robust statistical methods in simulation studies, a certain part of the data needs to be contaminated, so that the influence of these outliers on the robust estimators (and possibly their classical counterparts) can be studied. The term *contamination* is thereby used in a technical sense in this paper. In robust statistics, the distribution F of contaminated data is typically modeled as a mixture of distributions

$$F = (1-\varepsilon)G + \varepsilon H, \qquad (2.1)$$

where ε denotes the *contamination level*, G is the distribution of the non-contaminated part of the data and H is the distribution of the contamination (e.g., Maronna et al. 2006). Consequently, outliers may be modeled by a two-step process in simulation studies (Béguin and Hulliger 2008, Hulliger and Schoch 2009b):

1. Select the observations to be contaminated. The probabilities of selection may or may not depend on any other information in the data set.

2. Model the distribution of the outliers. The distribution may or may not depend on the original values of the selected observations.

A more detailed mathematical notation of this process with respect to the implementation in **simFrame** can be found in Alfons et al. (2010d).

Even though this is a rather simple concept, taking advantage of object-oriented programming techniques such as inheritance allows for a flexible implementation that can be extended by the user with custom contamination models. In **simFrame**, contamination is implemented based on control classes inheriting from `VirtualContControl`. For extensions of the framework, the user may define subclasses of this virtual class (see Section 2.7.3).

Figure 2.1 displays the full hierarchy of the available control classes for contamination. The basic virtual class contains the following slots:

target: A character vector defining the variables to be contaminated, or NULL to contaminate all variables (except the additional ones generated internally).

epsilon: A numeric vector giving the contamination levels to be used in the simulation.

With the contamination control classes available in **simFrame**, it is possible to contaminate whole groups (e.g., households) rather than individual observations. In addition, the probabilities for selecting items to be contaminated may depend on an auxiliary variable. In order to share these properties, another virtual class called ContControl is implemented. These are the additional slots:

grouping: A character string specifying a variable to be used for grouping.

aux: A character string specifying an auxiliary variable whose values are used as probability weights for selecting the items (observations or groups) to be contaminated.

The distribution of the contaminated data in simulation experiments may or may not depend on the original values. Similar to model-based data generation (see Section 2.4.1), the control class DCARContControl supports specifying a distribution function for generating the contamination. DCAR stands for distributed completely at random and corresponds to contamination independent of the original data. If a variable for grouping is specified, the same values are used for all observations in the same group. DCARContControl extends ContControl by the following slots:

distribution: A function for generating the data for the contamination, e.g., rmvnorm in package **mvtnorm** for a multivariate normal distribution.

dots: Additional arguments to be passed to distribution.

On the other hand, contamination based on the original values is realized by the control class DARContControl. DAR thereby stands for distributed at random. An arbitrary function may be used to modify the data. To do so, the original values of the observations to be contaminated are passed as its first argument. Thus the following slots are available in addition to those from ContControl:

fun: A function generating the values of the contaminated data based on the original values.

dots: Additional arguments to be passed to fun.

2. An object-oriented framework for statistical simulation

In the following example, a control object of class `DARContControl` is defined. The contamination level is set to 20% and the specified function multiplies the original values from variable "V2" of the observations to be contaminated by a factor 100.

```
R> cc <- DARContControl(target = "V2", epsilon = 0.2,
+     fun = function(x) x * 100)
```

If a control object for contamination is supplied, the framework calls the generic function `contaminate(x, control, ...)` in the simulation runs internally to add the contamination. In many applications, it is necessary to know which observations were contaminated, e.g., to evaluate outlier detection methods. Hence a logical variable, which is called ".contaminated" and indicates the contaminated observations, is added to the resulting data set. As an example, the data generated in Section 2.4.1 is contaminated below.

```
R> bar <- contaminate(foo, cc)
R> bar
```

```
            V1          V2 .contaminated
1  -1.289442540 -77.3344357          TRUE
2   0.009574644  -0.8925633         FALSE
3   0.846580448  -0.4691297         FALSE
4  -2.249086822  -0.5448488         FALSE
5   0.662837956   1.0378657         FALSE
6   0.460268469   0.0244493         FALSE
7  -0.687415163  -0.6423509         FALSE
8  -0.763840555  -1.0216259         FALSE
9  -0.761894738 -95.4736674          TRUE
10 -0.234446371   2.1031588         FALSE
```

Despite being designed for internal use in the simulation procedure, `contaminate()` also allows the control class to be specified as a character string (with `"DCARContControl"` being the default). In this case the slots may be supplied as arguments.

2.4.4 Insertion of missing values

Missing values are included in many data sets, in particular survey data hardly ever contain complete information. In practice, missing values often need to be imputed, which results in additional uncertainty in further statistical analysis (e.g., Little and Rubin 2002). This additional variability needs to be considered when computing variance estimates or confidence intervals. In simulation studies, it may therefore be of interest to study the properties of different imputation methods or to investigate the influence of missing values on point and variance estimates.

Three mechanisms generating missing values are commonly distinguished in the literature addressing missing data (e.g., Little and Rubin 2002):

- Missing completely at random (MCAR): The probability of missingness does not depend on any observed or missing information.

- Missing at random (MAR): The probability of missingness depends on the observed information.

- Missing not at random (MNAR): The probability of missingness depends on the missing information itself and may also depend on the observed information.

Similar to the implementation of the functionality for contamination, the insertion of missing data is handled by control classes extending `VirtualNAControl` (the hierarchy of the control classes is shown in Figure 2.1). This virtual class is the basis for extensions by the user (see Section 2.7.4). It consists of the following slots:

target: A character vector specifying the variables into which missing values should be inserted, or `NULL` to insert missing values into all variables (except the additional ones generated internally).

NArate: A numeric vector or matrix giving the missing value rates to be used in the simulation.

It should be noted that missing value rates may be selected individually for the target variables. The same missing value rates are used for all target variables if they are specified as a vector. If a matrix is supplied, on the other hand, the missing value rates to be used for each target variable are given by the respective column.

Extending `VirtualNAControl`, the control class `NAControl` allows whole groups to be set as missing rather than individual values. To account for MAR or MNAR situations instead of MCAR, an auxiliary variable of probability weights may be specified for each target variable. Furthermore, when studying robust methods for the analysis or imputation of incomplete data, it is sometimes desired to insert missing values only into non-contaminated observations. In other situations, a more realistic scenario in which missing values are also inserted into contaminated observations may be preferred. Both scenarios are implemented in the framework. These are the additional slots of `NAControl`:

grouping: A character string specifying a variable to be used for grouping.

aux: A character vector specifying auxiliary variables whose values are used as probability weights for selecting the values to be set as missing in the respective target variables.

intoContamination: A logical indicating whether missing values should also be inserted into contaminated observations. The default is to insert missings only into non-contaminated observations.

The following example shows how to define a control object of class `NAControl` that corresponds to an MCAR situation. For all variables, 30% of the values will be set as missing. However, missing values will only be inserted into non-contaminated observations.

```
R> nc <- NAControl(NArate = 0.3)
```

If a control object for missing data is supplied, the generic function `setNA(x, control, ...)` is called internally by the framework in the simulation runs to set the missing values. Below, missing values are inserted into the contaminated data from the previous section.

```
R> setNA(bar, nc)
```

	V1	V2	.contaminated
1	-1.2894425	-77.3344357	TRUE
2	NA	-0.8925633	FALSE
3	NA	-0.4691297	FALSE
4	NA	NA	FALSE
5	0.6628380	NA	FALSE
6	0.4602685	0.0244493	FALSE
7	-0.6874152	NA	FALSE
8	-0.7638406	-1.0216259	FALSE
9	-0.7618947	-95.4736674	TRUE
10	-0.2344464	2.1031588	FALSE

As `contaminate()`, the function `setNA()` is designed for internal use in the simulation procedure. Nevertheless, it is possible to supply the name of the control class as a character string (the default is `"NAControl"`), which allows the slots to be supplied as arguments.

2.4.5 Running simulations

The central component of the simulation framework is the generic function `runSimulation()`, which combines all the elements of the package into one convenient interface for running simulation studies. Based on a collection of control objects, it allows to perform even complex simulation experiments with just a few lines of code. Switching between simulation designs is possible with minimal programming effort as well, only some control objects need to be defined or modified. For design-based simulation, population data and a control object for sampling or previously set up samples may be passed to runSimulation(). For model-based simulation, on the other hand, a control object for data generation and the number of replications may be supplied.

In addition, the control class SimControl determines how the simulation runs are performed. These are the slots of SimControl (see also Figure 2.1):

contControl: A control object for contamination.

NAControl: A control object for inserting missing values.

design: A character vector specifying variables to be used for splitting the data into domains and performing the simulations on every domain.

fun: The function to be applied in the simulation runs.

dots: Additional arguments to be passed to fun.

SAE: A logical indicating whether small area estimation (see, e.g., Rao 2003) will be used in the simulation.

Most importantly, the function to be applied in the simulation runs needs to be defined. There are some requirements for the function:

- It must return a numeric vector, or a list with the two components values (a numeric vector) and add (additional results of any class, e.g., statistical models). Note that the latter is computationally slightly more expensive.

- A data.frame is passed to fun in every simulation run. The corresponding argument must be called x.

- If comparisons with the original data need to be made, e.g., for evaluating the quality of imputation methods, the function should have an argument called orig.

- If different domains are used in the simulation, the indices of the current domain can be passed to the function via an argument called domain.

One of the most important features of **simFrame** is that the supplied function is evaluated using try(). Therefore, if computations fail in one of the simulation runs, runSimulation() simply continues with the next run. The results from previous runs are not lost and the computation time has not been spent in vain.

Furthermore, control classes for adding contamination and missing values may be specified. In design-based simulations, contamination and nonresponse are added to the samples rather than the population, for maximum control over the amount of outliers or missing values (cf. Alfons et al. 2009). Another useful feature is that the data may be split into

2. An object-oriented framework for statistical simulation

different domains. The simulations, including contamination and the insertion of missing values, are then performed on every domain separately, unless small area estimation is used.

Concerning small area estimation, the following points have to be kept in mind. The design for splitting the data must be supplied and `SAE` must be set to `TRUE`. However, the data are not actually split into the specified domains. Instead, the whole data set is passed to the specified function. Also contamination and missing values are added to the whole data. Last, but not least, the function for the simulation runs must have a `domain` argument so that the current domain can be extracted from the whole data. In any case, small area estimation is not a main focus of the current version of **simFrame** and will therefore not be discussed further in this paper. Improving the support for small area estimation is future work.

For user convenience, the slots of the `SimControl` object may also be supplied as arguments. After running the simulations, the results of the individual simulation runs are combined and packed into an object of class `SimResults`. The most important slots are (see Figure 2.1 for a complete list):

values: A `data.frame` containing the simulation results.

add: A list containing additional simulation results, e.g., statistical models.

epsilon: The contamination levels used in the simulation.

NArate: The missing value rates used in the simulation.

seed: A list containing the seeds of the random number generator before and after the simulation, respectively.

An illustrative example for the use of `runSimulation()` is given in the following design-based simulation experiment. The synthetic EU-SILC example data of the package is thereby used as population data. It contains information about household income, but data is also available on the personal level (see Section 2.6.1 for more information on the data). From this data set, 50 samples of 500 persons are drawn with simple random sampling. In addition, the equivalized income of 2% of the sampled persons are multiplied by a factor 25. In every simulation run, the population mean income is estimated with the mean and the 2% trimmed mean of the sample. With the following commands, control objects for sampling and contamination are defined, along with the function for the simulation. Before calling `runSimulation()`, the seed of the random number generator is set for reproducibility of the results.

```
R> data("eusilcP")
R> sc <- SampleControl(size = 500, k = 50)
```

2.4 Implementation

```
R> cc <- DARContControl(target = "eqIncome", epsilon = 0.02,
+      fun = function(x) x * 25)
R> sim <- function(x) {
+      c(mean = mean(x$eqIncome), trimmed = mean(x$eqIncome, trim = 0.02))
+ }
R> set.seed(12345)
R> results <- runSimulation(eusilcP, sc, contControl = cc, fun = sim)
```

Methods for several frequently used generic functions are available to inspect the simulation results. Besides `head()`, `tail()` and `summary()` methods, a method for `aggregate()` is implemented. The latter can be used to calculate summary statistics of the results. By default, the mean is used as summary statistic. Depending on the simulation design, the summary statistics are are computed for different subsets of the results. These subsets are thereby given by the different combinations of contamination levels (if contamination is used), missing value rates (if missing values are inserted) and domains (if the simulations are performed on different domains of the data).

Below, the first parts of the simulation results are returned using `head()` and the average results are computed with `aggregate()`. For comparison, the true population mean is computed afterwards.

```
R> head(results)

  Run Sample Epsilon     mean trimmed
1   1      1    0.02 29609.10 20737.05
2   2      2    0.02 28834.87 20023.74
3   3      3    0.02 29819.41 20410.12
4   4      4    0.02 28840.16 20438.05
5   5      5    0.02 27323.11 19298.89
6   6      6    0.02 29614.79 20442.76

R> aggregate(results)

  Epsilon     mean trimmed
1    0.02 29697.64 20361.22

R> tv <- mean(eusilcP$eqIncome)
R> tv

[1] 20162.8
```

Various plots for simulation results are implemented in the framework, as discussed in the following section. In Figure 2.2, the results for this illustrative example are displayed by box plots and kernel density plots. The plots show the well-known fact that the mean is highly influenced by outliers. While the trimmed mean is not influenced by the contamination and has much smaller variance, there is still some bias. Since the outliers in this example are only in the upper tail of the data, the remaining bias results from trimming lower part as well.

Section 2.6 contains more elaborate examples for design-based and model-based simulation with detailed step-by-step instructions, as well as some motivation and interpretation.

2.4.6 Visualization

Visualization methods for the simulation results are based on **lattice** graphics (Sarkar 2008, 2010). If the simulation study has been divided into several domains, the results for each domain are displayed in a separate panel. Box plots and kernel density plots are implemented in the functions `simBwplot()` and `simDensityplot()`, respectively. For simulations involving different contamination levels or missing value rates, `simXyplot()` plots the average results against the contamination levels or missing value rates. In all of these plots, reference lines for the true values can be added. Moreover, the `plot()` method for the class `SimResults` selects a suitable graphical representation of the simulation results automatically.

Figure 2.2 shows the default plot and kernel density plots for the simulation results from the simple illustrative example in the previous section. Further examples for the visualization of simulation results are given in Section 2.6.

2.5 Parallel computing

Statistical simulation is *embarrassingly parallel*, hence computational performance can be increased by parallel computing. In **simFrame**, parallel computing is implemented using the R package **snow** (Rossini et al. 2007, Tierney et al. 2008, 2009), which is recommended by Schmidberger et al. (2009) in an analysis of the state-of-the-art in parallel computing with R. For setting up multiple samples and running simulations on a cluster, the functions `clusterSetup()` and `clusterRunSimulation()` are implemented. Note that all objects and packages required for the computations (including **simFrame**) need to be made available on every worker process. An example for parallel computing is presented in Section 2.6.3.

In order to ensure reproducibility of the simulation results, random number streams should be used. The R packages **rlecuyer** (L'Ecuyer et al. 2002, Sevcikova and Rossini 2009) and **rsprng** (Mascagni and Srinivasan 2000, Li 2010) for creating random number streams are supported by **snow** via the function `clusterSetupRNG()`. It should be noted

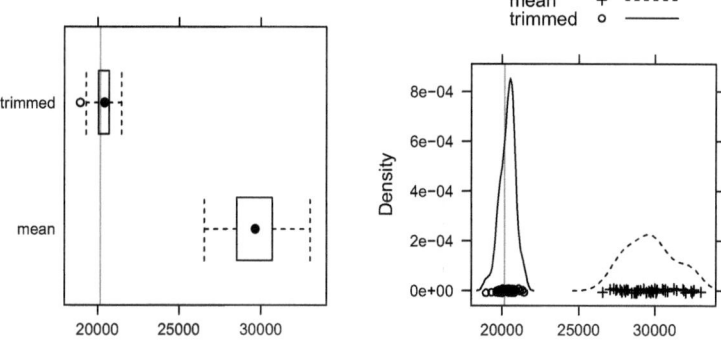

Figure 2.2: Simulation results from the simple illustrative example. *Left*: Default plot of results from a simulation study with one contamination level, in this example obtained by `plot(results, true = tv)`. *Right*: Kernel density plots of the simulation results, obtained by `simDensityplot(results, true = tv)`.

that the package **rstream** (L'Ecuyer and Leydold 2005, Leydold 2010) provides a faster connection to the C library by L'Ecuyer et al. (2002) than **rlecuyer**. Support in **simFrame** may thus be beneficial and may be added as a future task.

The R package **multicore** (Urbanek 2009) offers parallel computing on machines with multiple cores or CPUs. No data or code needs to be initialized and no additional R instances need to be started, hence spawning parallel processes is very fast. In addition, worker processes share memory and no network traffic is required, which allows fast computations on machines with a sufficient number of cores. However, **multicore** is not available for Microsoft Windows operating systems and it does not support random number streams out-of-the-box (as **snow** does). The incorporation of **multicore** into **simFrame** may be investigated in the future.

2.6 Using the framework

In this section, the use of **simFrame** is demonstrated on examples for design-based and model-based simulation. An example for parallel computing is included as well. Note that the only purpose of these examples is to illustrate the use of the package. It is not the aim of this section to provide a thorough analysis of the presented methodology, as this is beyond the scope of this paper.

2.6.1 Design-based simulation

The Laeken indicators are a set of indicators used to measure social cohesion in member states of the European Union and other European countries (cf. Atkinson et al. 2002). Most of the Laeken indicators are computed from EU-SILC (European Union Statistics on Income and Living Conditions) survey data. Synthetic EU-SILC data based on the Austrian sample from 2006 is included in **simFrame**. This data set was generated using the synthetic data generation framework by Alfons et al. (2010b) from package **simPopulation** (Alfons and Kraft 2010). It consists of 25 000 households with data available on the personal level, and is used as population data in this example. Note that this is an illustrative example, as the data set does not represent the true population sizes of Austria and its regions.

While only being a secondary Laeken indicator, the *Gini coefficient* is a frequently used measure of inequality and is widely studied in the literature. In the case of EU-SILC, the Gini coefficient is calculated based on an equivalized household income. In this example, the standard estimation method (EU-SILC 2004) is compared to two semiparametric approaches, which fit a Pareto distribution (e.g., Kleiber and Kotz 2003) to the upper tail of the data. Hill (1975) introduced the maximum-likelihood estimator, which is thus referred to as Hill estimator. The partial density component (PDC) estimator (Vandewalle et al. 2007), on the other hand, follows a robust approach. These methods are available in the R package **laeken** (Alfons et al. 2010a). A more detailed discussion on Pareto tail modeling with application to selected Laeken indicators can be found in Alfons et al. (2010e).

First, the required package and the data set need to be loaded. Furthermore, the seed of the random number generator is set for reproducibility of the results.

```
R> library("laeken")
R> data("eusilcP")
R> set.seed(12345)
```

Next, 100 samples of 1500 households are set up. Stratified sampling by regions combined with sampling of whole households rather than individuals can be achieved with one command.

```
R> set <- setup(eusilcP, design = "region", grouping = "hid",
+      size = c(75, 250, 250, 125, 200, 225, 125, 150, 100),
+      k = 100)
```

Since a robust method is going to be compared to two classical ones, a control object for contamination is defined. EU-SILC data typically contain a very low amount of outliers, therefore the equivalized household income of 0.5% of the households is contaminated. In addition, the contamination is generated by a normal distribution $\mathcal{N}(\mu, \sigma^2)$ with mean $\mu = 500\,000$ and standard deviation $\sigma = 10\,000$.

2.6 Using the framework

```
R> cc <- DCARContControl(target = "eqIncome", epsilon = 0.005,
+       grouping = "hid", dots = list(mean = 5e+05, sd = 10000))
```

The function for the simulation runs is quite simple as well. Its argument k determines the number of households whose income is modeled by a Pareto distribution.

```
R> sim <- function(x, k) {
+       g <- gini(x$eqIncome, x$.weight)$value
+       eqIncHill <- fitPareto(x$eqIncome, k = k, method = "thetaHill",
+           groups = x$hid)
+       gHill <- gini(eqIncHill, x$.weight)$value
+       eqIncPDC <- fitPareto(x$eqIncome, k = k, method = "thetaPDC",
+           groups = x$hid)
+       gPDC <- gini(eqIncPDC, x$.weight)$value
+       c(standard = g, Hill = gHill, PDC = gPDC)
+ }
```

With all necessary objects available, running the simulation experiment is only one more command. Note that simulations are performed separately for each gender. The value of k for the Pareto distribution is thereby set to 125.

```
R> results <- runSimulation(eusilcP, set, contControl = cc,
+       design = "gender", fun = sim, k = 125)
```

The head() and aggregate() methods are used to take a look at the simulation results. In this case, aggregate() computes the average results for each subset.

```
R> head(results)
```

	Run	Sample	Epsilon	gender	standard	Hill	PDC
1	1	1	0.005	male	34.58446	29.96658	26.61415
2	1	1	0.005	female	38.82356	33.93700	28.82045
3	2	2	0.005	male	34.34853	29.09325	27.66380
4	2	2	0.005	female	36.38429	30.06097	27.42663
5	3	3	0.005	male	33.39992	30.54211	23.96698
6	3	3	0.005	female	35.12883	30.51336	26.06518

```
R> aggregate(results)
```

	Epsilon	gender	standard	Hill	PDC
1	0.005	male	33.18580	29.00265	26.21119
2	0.005	female	35.61341	31.28984	27.69054

2. An object-oriented framework for statistical simulation

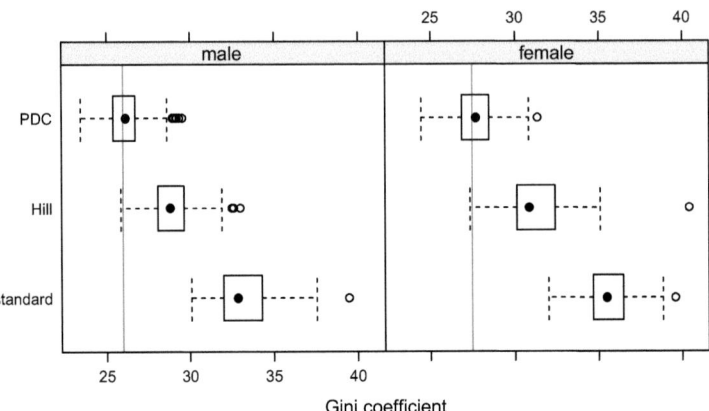

Figure 2.3: Default plot of results from a simulation study with one contamination level and different domains, in this example obtained by plot(results, true = tv, xlab = "Gini coefficient").

For comparison with the simulation results, the true values of the Gini coefficient need to be computed. These can be added as reference lines to the plot of the simulation results (see Figure 2.3).

```
R> tv <- simSapply(eusilcP, "gender", function(x) gini(x$eqIncome)$value)
```

Figure 2.3 shows that even a small proportion of outliers completely corrupts the standard estimation of the Gini coefficient. Also fitting the Pareto distribution with the Hill estimator is highly influenced by contamination, whereas the robust PDC estimator leads to excellent results. But most importantly, this example shows that even complex simulation designs require only a few lines of code.

Further examples for design-based simulation that demonstrate the strengths of the framework can be found in a supplementary paper. This supplementary paper is also included in **simFrame** as a package vignette (Leisch 2003).

2.6.2 Model-based simulation

In this section, model-based simulation is demonstrated using an example for compositional data. An observation $\boldsymbol{x} = (x_1, \ldots, x_D)$ is by definition a *D-part composition* if, and only if, $x_i > 0$, $i = 1, \ldots, D$, and all relevant information is contained in the ratios between the components (Aitchison 1986). Consequently, compositional data contain only relative information. The information is essentially the same if an observation is multiplied with a positive constant. But if the value of one component changes, the other components need to change accordingly. Examples for compositional data are element concentrations in chemical analysis of a sample material or monthly household expenditures on different spending categories such as housing, food or leisure activities.

It is important to note that compositional data have no direct representation in the Euclidean space and that their geometry is entirely different (see Aitchison 1986). The sample space of D-part compositions is called the *simplex* and a suitable distance measure is the *Aitchison distance* d_A (Aitchison 1992, Aitchison et al. 2000). Fortunately, there exists an isometric transformation from the D-dimensional simplex to $\mathbb{R}^{D-1}$, which is called the *isometric logratio* (ilr) transformation (Egozcue et al. 2003). With this transformation, the Aitchison distance can be expressed as

$$d_A(\boldsymbol{x},\boldsymbol{y}) = d_E(\mathrm{ilr}(\boldsymbol{x}),\mathrm{ilr}(\boldsymbol{y})), \tag{2.2}$$

where d_E denotes the Euclidean distance.

Hron et al. (2010) introduced imputation methods for compositional data, which are implemented in the R package **robCompositions** (Templ et al. 2009, 2010b). While the package is focused on robust methods, only classical imputation methods are used in this example. The first method is a modification of k-nearest neighbor ($k$nn) imputation (Troyanskaya et al. 2001), the second follows an iterative model-based approach using least squares (LS) regression.

Before any computations are performed, the required packages are loaded and the seed of the random number generator is set for reproducibility.

```
R> library("robCompositions")
R> library("mvtnorm")
R> set.seed(12345)
```

The data in this example are generated by a *normal distribution on the simplex*, denoted by $\mathcal{N}_\mathcal{S}^D(\boldsymbol{\mu},\boldsymbol{\Sigma})$ (e.g., Mateu-Figueras et al. 2008). A random composition $\boldsymbol{x} = (x_1,\ldots,x_D)$ follows this distribution if, and only if, the vector of ilr transformed variables follows a multivariate normal distribution on $\mathbb{R}^{D-1}$ with mean vector $\boldsymbol{\mu}$ and covariance matrix $\boldsymbol{\Sigma}$. The following commands create a control object for generating 150 realizations of a random variable $\boldsymbol{X} \sim \mathcal{N}_\mathcal{S}^4(\boldsymbol{\mu},\boldsymbol{\Sigma})$ with

$$\boldsymbol{\mu} = \begin{pmatrix} 0 \\ 2 \\ 3 \end{pmatrix} \quad \text{and} \quad \boldsymbol{\Sigma} = \begin{pmatrix} 1 & -0.5 & 1.4 \\ -0.5 & 1 & -0.6 \\ 1.4 & -0.6 & 2 \end{pmatrix}.$$

2. An object-oriented framework for statistical simulation

```
R> crnorm <- function(n, mean, sigma) invilr(rmvnorm(n, mean, sigma))
R> sigma <- matrix(c(1, -0.5, 1.4, -0.5, 1, -0.6, 1.4, -0.6, 2), 3, 3)
R> dc <- DataControl(size = 150, distribution = crnorm,
+      dots = list(mean = c(0, 2, 3), sigma = sigma))
```

Furthermore, a control object for inserting missing values needs to be created. In every variable, 5% of the observations are set as missing completely at random.

```
R> nc <- NAControl(NArate = 0.05)
```

For the two selected imputation methods, the *relative Aitchison distance* between the original and the imputed data (cf. the simulation study in Hron et al. 2010) is computed in every simulation run.

```
R> sim <- function(x, orig) {
+      i <- apply(x, 1, function(x) any(is.na(x)))
+      ni <- length(which(i))
+      xKNNa <- impKNNa(x)$xImp
+      xLS <- impCoda(x, method = "lm")$xImp
+      c(knn = aDist(xKNNa, orig)/ni, LS = aDist(xLS, orig)/ni)
+ }
```

The simulation can then be run with the following command:

```
R> results <- runSimulation(dc, nrep = 100, NAControl = nc,
+      fun = sim)
```

As in the previous example, the results are inspected using head() and aggregate().

```
R> head(results)
```

	Run	Rep	NArate	knn	LS
1	1	1	0.05	0.3438037	0.2880082
2	2	2	0.05	0.2812875	0.1792722
3	3	3	0.05	0.4211458	0.2680063
4	4	4	0.05	0.3004898	0.2259575
5	5	5	0.05	0.3317538	0.2011256
6	6	6	0.05	0.4174775	0.3070453

```
R> aggregate(results)
```

	NArate	knn	LS
1	0.05	0.4379644	0.3007417

2.6 Using the framework

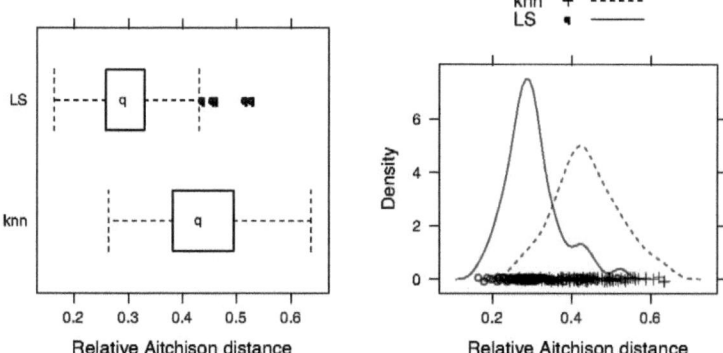

Figure 2.4: *Left*: Default plot of results from a simulation study with one missing value rate, in this example obtained by plot(results, xlab = "Relative Aitchison distance"). *Right*: Kernel density plots of the simulation results, obtained by simDensityplot(results, alpha = 0.6, xlab = "Relative Aitchison distance").

Box plots and kernel density plots of the simulation results are presented in Figure 2.4. Since the imputation methods in this example are evaluated in terms of a relative distance measure, values closer to 0 indicate better performance. Clearly, the iterative model-based procedure leads to better results than the modified *k*nn approach with respect to the relative Aitchison distance. This is not a surprising result, as the latter is used as a starting point in the iterative procedure. For serious evaluation of the imputation methods, however, also other criteria need to be taken into account (e.g., how well the variability of the multivariate data is reflected; see Hron et al. 2010).

2.6.3 Parallel computing

Using parallel computing, computation time may be significantly decreased in statistical simulation. In this section, the example for model-based simulation from before is extended to more than one missing value rate. Hence some of the objects are already defined above, but in order to provide a complete description on how to perform parallel computing with **simFrame**, these definitions are repeated here.

The first step is to start a **snow** cluster. In this example, four parallel worker processes on the local machine are initialized.

```
R> cl <- makeCluster(4, type = "SOCK")
```

2. An object-oriented framework for statistical simulation

All the functions and packages required for the computations (including **simFrame**) need to be loaded on the worker processes.

```
R> clusterEvalQ(cl, {
+       library("simFrame")
+       library("robCompositions")
+       library("mvtnorm")
+ })
```

For reproducibility of the results, a random number stream is generated.

```
R> clusterSetupRNG(cl, seed = 12345)
```

Control objects for data generation and the insertion of missing values, as well as the function for the simulation runs are defined as in the previous section. The only difference is that multiple missing value rates (1%, 3%, 5%, 7% and 9%) are used in this example.

```
R> crnorm <- function(n, mean, sigma) invilr(rmvnorm(n, mean, sigma))
R> sigma <- matrix(c(1, -0.5, 1.4, -0.5, 1, -0.6, 1.4, -0.6, 2), 3, 3)
R> dc <- DataControl(size = 150, distribution = crnorm,
+       dots = list(mean = c(0, 2, 3), sigma = sigma))
R> nc <- NAControl(NArate = c(0.01, 0.03, 0.05, 0.07, 0.09))
R> sim <- function(x, orig) {
+       i <- apply(x, 1, function(x) any(is.na(x)))
+       ni <- length(which(i))
+       xKNNa <- impKNNa(x)$xImp
+       xLS <- impCoda(x, method = "lm")$xImp
+       c(knn = aDist(xKNNa, orig)/ni, LS = aDist(xLS, orig)/ni)
+ }
```

These objects need to be made available on the worker processes. Since they are small in size, they are exported. Note that large objects, e.g., data sets for design-based simulation, should rather be constructed on the worker processes, as computation is much faster than network communication (Schmidberger et al. 2009).

```
R> clusterExport(cl, c("crnorm", "sigma", "dc", "nc", "sim"))
```

Then only one more command is needed to run the simulation.

```
R> results <- clusterRunSimulation(cl, dc, nrep = 100, NAControl = nc,
+       fun = sim)
```

Last, the cluster needs to be stopped after carrying out the simulation study in order to ensure that the worker processes are properly shut down.

```
R> stopCluster(cl)
```

After the parallel computations have finished, the simulation results can be inspected as usual. In this example, the `aggregate()` method returns the average results of the relative distances for each missing value rate.

```
R> head(results)

  Run Rep NArate      knn        LS
1   1   1   0.01 0.3651501 0.2572328
2   2   1   0.03 0.4699935 0.3629816
3   3   1   0.05 0.3950942 0.1858227
4   4   1   0.07 0.4145427 0.4004828
5   5   1   0.09 0.4455230 0.3474315
6   6   2   0.01 0.3189354 0.1451337

R> aggregate(results)

  NArate       knn        LS
1   0.01 0.3869185 0.2553123
2   0.03 0.4169886 0.2902240
3   0.05 0.4480634 0.3064850
4   0.07 0.4865589 0.3442564
5   0.09 0.5024330 0.3568477
```

Figure 2.5 visualizes the simulation results. On the left hand side, the average relative Aitchison distances are plotted against the missing value rates. On the right hand side, kernel density plots for a specified missing value rate (7%) is shown. The results are not much different from those in the previous section. Note that the difference of the average results for the two methods remains quite constant in this simulation example.

2.7 Extending the framework

One of the main advantages of the S4 implementation of **simFrame** is that it provides clear interfaces for user-defined extensions. With the available control classes for data generation, sampling, contamination and the insertion of missing data, the framework is highly flexible and can be used for a wide range of simulation designs. Nevertheless, extensions may sometimes be desired for specialized functionality. In order to extend the framework, developers can implement custom control classes and the corresponding methods.

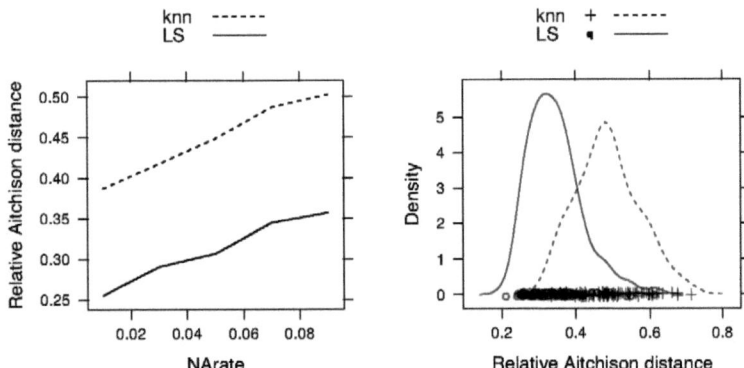

Figure 2.5: *Left*: Default plot of results from a simulation study with multiple missing value rates, in this example obtained by `plot(results, ylab = "Relative Aitchison distance")`. *Right*: Kernel density plots of the simulation results for a specified missing value rate (7%), obtained by `simDensityplot(results, NArate=0.07, alpha = 0.6, xlab = "Relative Aitchison distance")`.

2.7.1 Model-based data

The control class `DataControl` available in simFrame is quite simple but general. For user-defined data generation models, it often suffices to implement a function and use it as the `distribution` slot in the `DataControl` object. This function should have the number of observations to be generated as its first argument, as illustrated in the code skeleton in Figure 2.6 (*top*). The name of the argument is thereby not important. Furthermore, the function should return an object that can be coerced to a `data.frame`.

However, if more specialized data generation models are required, the framework can be extended by defining a control class extending `VirtualDataControl` and the corresponding method for the generic function `generate()`. If, e.g., a specific distribution or mixture of distributions is frequently used in simulation experiments, a distinct control class may be more convenient for the user. Figure 2.6 (*bottom*) contains the code skeleton for such an extension.

2.7.2 Sampling

In **simFrame**, the control class `SampleControl` is highly flexible and allows stratified sampling as well as sampling of whole groups rather than individuals with a specified sampling method. Hence it is often sufficient to implement the desired sampling method for the simple non-stratified case to extend the existing framework. However, there are some restrictions

2.7 Extending the framework

```
myDataGeneration <- function(size, ...) {
    # computations
}
```

```
setClass("MyDataControl",
    # class definition
    contains = "VirtualDataControl")

setMethod("generate",
    signature(control = "MyDataControl"),
    function(control) {
        # method definition
    })
```

Figure 2.6: *Top:* Code skeleton for a user-defined data generation method. *Bottom:* Code skeleton for extending model-based data generation with a custom control class and the corresponding method for `generate()`.

```
myPoisson <- function(prob) {
    require(sampling)
    which(as.logical(UPpoisson(prob)))
}
```

```
setClass("MySampleControl",
    # definition of additional properties
    contains = "VirtualSampleControl")

setMethod("setup",
    signature(x = "data.frame", control = "MySampleControl"),
    function(x, control) {
        # method definition
    })
setMethod("clusterSetup",
    signature(x = "data.frame", control = "MySampleControl"),
    function(cl, x, control) {
        # method definition
    })
```

Figure 2.7: *Top:* User-defined function for Poisson sampling. *Bottom:* Code skeleton for user-defined setup of multiple samples with a custom control class and the corresponding methods for `setup()` and `clusterSetup()`.

on the argument names of the function, which should return a vector containing the indices of the sampled observations.

- If the sampling method needs population data as input, the corresponding argument should be called `x` and should expect a `data.frame`.

- If it only needs the population size as input, the argument should be called `N`.

2. An object-oriented framework for statistical simulation

- If necessary, the argument for the sample size should be called `size`.

- If necessary, the argument for the probability weights should be called `prob`.

Note that the function is not expected to have both x and N as arguments, and that the latter is much faster for stratified sampling or group sampling. Furthermore, a function with `prob` as its only argument is perfectly valid (for probability proportional to size sampling). Figure 2.7 (*top*) shows an example for Poisson sampling using the implementation in package **sampling** (Tillé and Matei 2009).

Nevertheless, for very complex sampling procedures, it is possible to define a control class extending `VirtualSampleControl` and the corresponding `setup()` method. The code skeleton for such an extension is shown in Figure 2.7 (*bottom*). In order to optimize computational performance, it is necessary to efficiently set up multiple samples. Thereby the slot k of `VirtualSampleControl` needs to be used to control the number of samples, and the resulting object must be of class `SampleSetup`. For using parallel computing to set up samples with a self-defined control class, a method for `clusterSetup()` may be defined.

2.7.3 Contamination

A wide range of contamination models is covered by the control classes `DCARContControl` and `DARContControl`. However, other contamination models can be added by defining a control class inheriting from `VirtualContControl` and the corresponding method for `contaminate()` (see the code skeleton in Figure 2.8). Note that `VirtualContControl` contains the slots `target` and `epsilon` for selecting the target variable(s) and contamination level(s), respectively. In case the contaminated observations need to be identified at a later stage of the simulation, e.g., if conflicts with inserting missing values should be avoided, a logical indicator variable `".contaminated"` should be added to the returned data set.

```
setClass("MyContControl",
    # definition of additional properties
    contains = "VirtualContControl")

setMethod("contaminate",
    signature(x = "data.frame", control = "MyContControl"),
    function(x, control, i) {
        # method definition
    })
```

Figure 2.8: Code skeleton for a user-defined control class for contamination and the corresponding method for `contamintate()`.

2.7.4 Insertion of missing values

Similar to extending the framework for model-based data generation and contamination, user-defined missing value models can be added by defining a control class extending the virtual class `VirtualNAControl` and the corresponding method for the generic function `setNA()` (see the code skeleton in Figure 2.9). The slots `target` and `NArate` for selecting the target variable(s) and missing value rate(s), respectively, are inherited from `VirtualNAControl`.

```
setClass("MyNAControl",
    # definition of additional properties
    contains = "VirtualNAControl")
setMethod("setNA",
    signature(x = "data.frame", control = "MyNAControl"),
    function(x, control, i) {
        # method definition
    })
```

Figure 2.9: Code skeleton for a user-defined control class for the insertion of missing values and the corresponding method for `setNA()`.

2.8 Conclusions and outlook

The flexible, object-oriented implementation of **simFrame** allows researchers to make use of a wide range of simulation designs with a minimal effort of programming. Control classes are used to handle data generation, sampling, contamination and the insertion of missing values. Due to the use of control objects, switching from one simulation design to another requires only minimal programming effort. Developers can easily extend the existing framework with user-defined classes and methods. Guidelines for simulation studies in research projects can therefore be established by selecting or implementing control classes and agreeing upon parameter values, thus ensuring comparable results from different researchers. Based on the structure of the simulation results, an appropriate plot method is selected automatically. Hence **simFrame** is widely applicable for gaining insight into the quality of statistical methods. Furthermore, since the workload in statistical simulation is embarrassingly parallel, **simFrame** supports parallel computing using **snow** to increase computational performance.

Future plans include to further develop the model-based data generation facilities and implement mixed simulation designs, to improve the support for small area estimation, as well as to extend the framework with different sampling methods and more specialized contamination and missing data models. In addition, adding support of additional packages for parallel computing and random number streams may be considered. Concerning large data sets, the incorporation of the package **ff** for memory-efficient storage may be investigated.

2. An object-oriented framework for statistical simulation

Computational details All computations in this paper were performed using **Sweave** (Leisch 2002a,b) with R version 2.12.0 and **simFrame** version 0.3.6. The most recent version of the package is always available from CRAN (the Comprehensive R Archive Network, http://cran.R-project.org), and (a slightly modified and up-to-date version of) this paper is also included as a package vignette (Leisch 2003).

Acknowledgments This work was partly funded by the European Union (represented by the European Commission) within the 7th framework programme for research (Theme 8, Socio-Economic Sciences and Humanities, Project AMELI (Advanced Methodology for European Laeken Indicators), Grant Agreement No. 217322). Visit http://ameli.surveystatistics.net for more information on the project. Furthermore, we would like to thank two anonymous referees for their constructive remarks that helped to improve the package and the paper.

Chapter 3

Applications of statistical simulation in the case of EU-SILC: Using the R package simFrame

Supplementary material to Alfons et al. (2010c) (Chapter 2), which has been accepted for publication in the *Journal of Statistical Software*.

Andreas Alfons[a], Matthias Templ[a,b], Peter Filzmoser[a]

[a] Department of Statistics and Probability Theory, Vienna University of Technology
[b] Methods Unit, Statistics Austria

Abstract This paper demonstrates the use of **simFrame** for various simulation designs in a practical application with EU-SILC data. It presents the full functionality of the framework regarding sampling designs, contamination models, missing data mechanisms and performing simulations separately on different domains. Due to the use of control objects, switching from one simulation design to another requires only minimal changes in the code. Using bespoke R code, on the other hand, changing the code to switch between simulation designs would require much greater effort. Furthermore, parallel computing with **simFrame** is demonstrated.

Keywords R, statistical simulation, EU-SILC

3.1 Introduction

This is a supplementary paper to "An Object-Oriented Framework for Statistical Simulation: The R Package **simFrame**" (Alfons et al. 2010c) and demonstrates the use of **simFrame** (Alfons 2010) in R (R Development Core Team 2010) for various simulation designs in a practical application. It extends the example for design-based simulation in Alfons et al. (2010c) (Example 6.1). Different simulation designs in terms of sampling, contamination and missing data are thereby investigated to present the strengths of the framework.

Note that the paper is supplementary material and is supposed to be read after studying Alfons et al. (2010c). It does not give a detailed discussion about the motivation for the framework, nor does it describe the design or implementation of the package. Instead it is focused on showing its full functionality for design-based simulation in additional code examples with brief explanations. However, model-based simulation is not considered here.

The European Union Statistics on Income and Living Conditions (EU-SILC) is panel survey conducted in EU member states and other European countries and serves as basis for measuring risk-of-poverty and social cohesion in Europe. An important indicator calculated from this survey is the *Gini coefficient*, which is a well-known measure of inequality. In the following examples, the standard estimation method (EU-SILC 2004) is compared to two semiparametric methods under different simulation designs. The two semiparametric approaches are based on fitting a Pareto distribution (e.g., Kleiber and Kotz 2003) to the upper tail of the data. In the first approach, the classical Hill estimator (Hill 1975) is used to estimate the shape parameter of the Pareto distribution, while the second uses the robust partial density component (PDC) estimator (Vandewalle et al. 2007). All these methods are implemented in the R package **laeken** (Alfons et al. 2010a). For a more detailed discussion on Pareto tail modeling in the case of the Gini coefficient and a related measure of inequality, the reader is referred to Alfons et al. (2010e).

The example data set of **simFrame** is used as population data throughout the paper. It consists of 58 654 observations from 25 000 households and was synthetically generated from Austrian EU-SILC survey data from 2006 using the data simulation methodology by Alfons et al. (2010b), which is implemented R package **simPopulation** (Alfons and Kraft 2010).

3.2 Application of different simulation designs to EU-SILC

First, the required packages and the data set need to be loaded.

```
R> library("simFrame")
R> bwtheme <- canonical.theme(color = FALSE)
R> bwtheme$superpose.line$lty <- c(3:1, 4:7)
R> lattice.options(default.theme = bwtheme)
R> library("laeken")
R> data("eusilcP")
```

3.2 Application of different simulation designs to EU-SILC

Then, the function to be run in every iteration is defined. Its argument k determines the number of households whose income is modeled by a Pareto distribution. Since the Gini coefficient is calculated based on an equivalized household income, all individuals of a household in the upper tail receive the same value.

```
R> sim <- function(x, k) {
+     x <- x[!is.na(x$eqIncome), ]
+     g <- gini(x$eqIncome, x$.weight)$value
+     eqIncHill <- fitPareto(x$eqIncome, k = k, method = "thetaHill",
+         groups = x$hid)
+     gHill <- gini(eqIncHill, x$.weight)$value
+     eqIncPDC <- fitPareto(x$eqIncome, k = k, method = "thetaPDC",
+         groups = x$hid)
+     gPDC <- gini(eqIncPDC, x$.weight)$value
+     c(standard = g, Hill = gHill, PDC = gPDC)
+ }
```

This function is used in the following examples, which are designed to exhibit the strengths of the framework. In order to change from one simulation design to another, all there is to do is to define or modify control objects and supply them to the function runSimulation().

3.2.1 Basic simulation design

In this basic simulation design, 100 samples of 1500 households are drawn using simple random sampling. Note that the setup() function is not used to permanently store the samples in an object. This is simply not necessary, since the population is rather small and the sampling method is straightforward. Furthermore, the Pareto distribution is fitted to the 175 households with the largest equivalized income.

```
R> set.seed(12345)
R> sc <- SampleControl(grouping = "hid", size = 1500, k = 100)
R> results <- runSimulation(eusilcP, sc, fun = sim, k = 175)
```

In order to inspect the simulation results, methods for several frequently used generic functions are implemented. Besides head(), tail() and summary() methods, a method for computing summary statistics with aggregate() is available. By default, the mean is used as summary statistic. Moreover, the plot() method selects a suitable graphical representation of the simulation results automatically. A reference line for the true value can thereby be added as well.

```
R> head(results)
```

3. Applications of statistical simulation in the case of EU-SILC

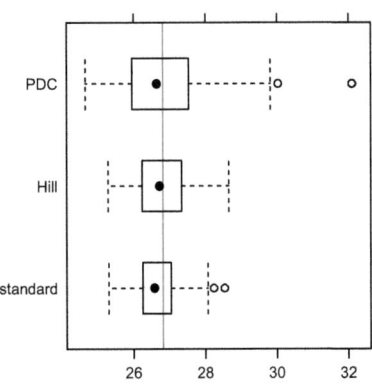

Figure 3.1: Simulation results for the basic simulation design.

```
  Run Sample standard     Hill      PDC
1   1      1 26.56793 26.48025 25.66614
2   2      2 26.98203 27.73124 26.39318
3   3      3 27.07081 27.11886 25.52524
4   4      4 26.86841 27.70216 25.71355
5   5      5 26.43215 26.49267 25.64191
6   6      6 26.96175 27.13876 27.17536

R> aggregate(results)

standard     Hill      PDC
26.65621 26.79016 26.89564

R> tv <- gini(eusilcP$eqIncome)$value
R> plot(results, true = tv)
```

Figure 3.1 shows the resulting box plots of the simulation results for the basic simulation design. While the PDC estimator comes with larger variability, all three methods are on average quite close to the true population value. This is also an indication that the choice of the number of households for fitting the Pareto distribution is suitable.

3.2.2 Using stratified sampling

The most frequently used sampling designs in official statistics are implemented in **simFrame**. In order to switch to another sampling design, only the corresponding control object needs to be changed. In this example, stratified sampling by region is performed. The sample sizes for the different strata are specified by using a vector for the slot `size` of the control object.

```
R> set.seed(12345)
R> sc <- SampleControl(design = "region", grouping = "hid",
+       size = c(75, 250, 250, 125, 200, 225, 125, 150, 100),
+       k = 100)
R> results <- runSimulation(eusilcP, sc, fun = sim, k = 175)
```

As before, the simulation results are inspected with `head()` and `aggregate()`. A plot of the simulation results is produced as well.

```
R> head(results)
```

```
  Run Sample standard    Hill      PDC
1   1      1 1 27.08652 27.22293 27.66753
2   2      2 2 26.80670 27.35874 25.93378
3   3      3 3 26.68113 27.03964 26.60062
4   4      4 4 25.84734 26.52346 25.18298
5   5      5 5 26.05449 26.26848 26.60331
6   6      6 6 26.98439 27.01396 26.48090
```

```
R> aggregate(results)
```

```
standard     Hill      PDC
26.71792 26.85375 26.86248
```

```
R> tv <- gini(eusilcP$eqIncome)$value
R> plot(results, true = tv)
```

Figure 3.2 contains the plot of the simulation results for the simulation design with stratified sampling. The results are very similar to those from the basic simulation design with simple random sampling. On average, all three investigated methods are quite close to the true population value.

3.2.3 Adding contamination

When evaluating robust methods in simulation studies, contamination needs to be added to the data to study the influence of these outliers on the robust estimators and their classical counterparts. In **simFrame**, contamination is specified by defining a control object. Various contamination models are thereby implemented in the framework. Keep in mind that the

3. Applications of statistical simulation in the case of EU-SILC

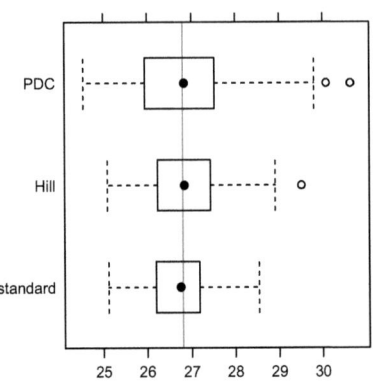

Figure 3.2: Simulation results for the simulation design with stratified sampling.

term *contamination* is used in a technical sense here (see Alfons et al. 2010c,d, for an exact definition) and that contamination is modeled as a two step process (see also Béguin and Hulliger 2008, Hulliger and Schoch 2009b). In this example, 0.5% of the households are selected to be contaminated using simple random sampling. The equivalized income of the selected households is then drawn from a normal distribution with mean $\mu = 500\,000$ and standard deviation $\sigma = 10\,000$.

```
R> set.seed(12345)
R> sc <- SampleControl(design = "region", grouping = "hid",
+      size = c(75, 250, 250, 125, 200, 225, 125, 150, 100),
+      k = 100)
R> cc <- DCARContControl(target = "eqIncome", epsilon = 0.005,
+      grouping = "hid", dots = list(mean = 5e+05, sd = 10000))
R> results <- runSimulation(eusilcP, sc, contControl = cc,
+      fun = sim, k = 175)
```

The head(), aggregate() and plot() methods are again used to take a look at the simulation results. Note that a column is added that indicates the contamination level used.

```
R> head(results)
```

	Run	Sample	Epsilon	standard	Hill	PDC
1	1	1	0.005	32.71453	29.12110	27.03731
2	2	2	0.005	34.22065	31.62709	26.24857

3.2 Application of different simulation designs to EU-SILC

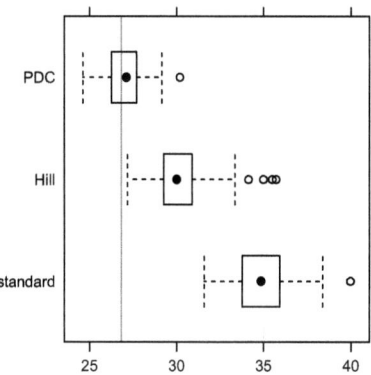

Figure 3.3: Simulation results for the simulation design with stratified sampling and contamination.

```
3   3     3   0.005 33.56878 28.49760 28.00937
4   4     4   0.005 35.26346 29.57160 26.25621
5   5     5   0.005 33.79720 29.15945 25.61514
6   6     6   0.005 34.72069 28.58610 27.22342

R> aggregate(results)

  Epsilon standard    Hill      PDC
1   0.005 34.88922 30.26179 27.02093

R> tv <- gini(eusilcP$eqIncome)$value
R> plot(results, true = tv)
```

In Figure 3.3, the resulting box plots are presented. The figure shows that such a small amount of contamination is enough to completely corrupt the standard estimation of the Gini coefficient. Using the classical Hill estimator to fit the Pareto distribution is still highly influenced by the outliers, whereas the PDC estimator leads to very accurate results.

3.2.4 Performing simulations separately on different domains

Data sets from official statistics typically contain strong heterogeneities, therefore indicators are usually computed for subsets of the data as well. Hence it is often of interest to investigate the behavior of indicators on different subsets in simulation studies. In **simFrame**, this can

3. Applications of statistical simulation in the case of EU-SILC

be done by simply specifying the `design` argument of the function `runSimulation()`. In the case of extending the example from the previous section, the framework then splits the samples, inserts contamination into each subset and calls the supplied function for these subsets automatically. With bespoke R code, the user would need to take care of this with a loop-like structure such as a `for` loop or a function from the `apply` family.

In the following example, the simulations are performed separately for each gender. It should be noted that the value of k for the Pareto distribution is thus changed to 125. This is the same as Example 6.1 from Alfons et al. (2010c), except that a control object for sampling is supplied to `runSimulation()` instead of setting up the samples beforehand and storing them in an object.

```
R> set.seed(12345)
R> sc <- SampleControl(design = "region", grouping = "hid",
+      size = c(75, 250, 250, 125, 200, 225, 125, 150, 100),
+      k = 100)
R> cc <- DCARContControl(target = "eqIncome", epsilon = 0.005,
+      grouping = "hid", dots = list(mean = 5e+05, sd = 10000))
R> results <- runSimulation(eusilcP, sc, contControl = cc,
+      design = "gender", fun = sim, k = 125)
```

Below, the results are inspected using `head()` and `aggregate()`. The `aggregate()` method thereby computes the summary statistic for each subset automatically. Also the `plot()` method displays the results for the different subsets in different panels by taking advantage of the **lattice** system (Sarkar 2008, 2010). In order to compute the true values for each subset, the function `simSapply()` is used.

```
R> head(results)

  Run Sample Epsilon gender standard     Hill      PDC
1   1      1   0.005   male 34.58446 29.96658 26.61415
2   1      1   0.005 female 38.82356 33.93700 28.82045
3   2      2   0.005   male 34.34853 29.09325 27.66380
4   2      2   0.005 female 36.38429 30.06097 27.42663
5   3      3   0.005   male 33.39992 30.54211 23.96698
6   3      3   0.005 female 35.12883 30.51336 26.06518

R> aggregate(results)

  Epsilon gender standard     Hill      PDC
1   0.005   male 33.18580 29.00265 26.21119
2   0.005 female 35.61341 31.28984 27.69054
```

3.2 Application of different simulation designs to EU-SILC

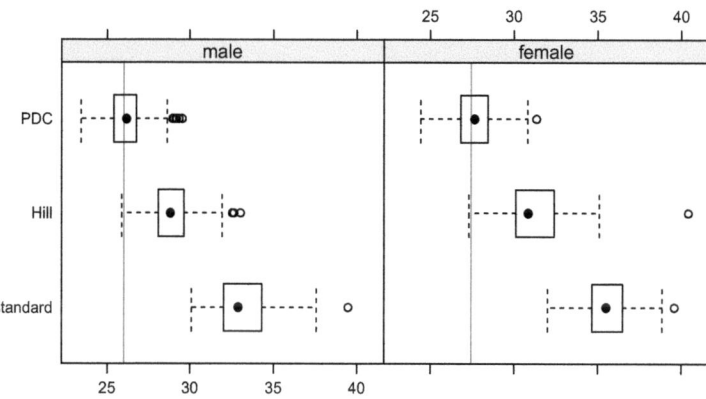

Figure 3.4: Simulation results for the simulation design with stratified sampling, contamination and performing the simulations separately for each gender.

```
R> tv <- simSapply(eusilcP, "gender", function(x) gini(x$eqIncome)$value)
R> plot(results, true = tv)
```

The resulting plots are shown in Figure 3.4, which is the same as Figure 2 in Alfons et al. (2010c). Clearly, the PDC estimator leads to excellent results for both subsets, while the two classical approaches are in both cases highly influenced by the outliers.

3.2.5 Using multiple contamination levels

To get a more complete picture of the behavior of robust methods, more than one level of contamination is typically investigated in simulation studies. The only necessary modification of the code is to use a vector of contamination levels as the slot epsilon of the contamination control object. In this example, the contamination level is varied from 0% to 1% in steps of 0.25%. With bespoke R code, the user would have to add another loop-like structure to the code and collect the results in a suitable data structure. In **simFrame**, this is handled internally by the framework.

```
R> set.seed(12345)
R> sc <- SampleControl(design = "region", grouping = "hid",
+      size = c(75, 250, 250, 125, 200, 225, 125, 150, 100),
+      k = 100)
R> cc <- DCARContControl(target = "eqIncome", epsilon = c(0,
```

3. Applications of statistical simulation in the case of EU-SILC

```
+         0.0025, 0.005, 0.0075, 0.01), dots = list(mean = 5e+05,
+         sd = 10000))
R> results <- runSimulation(eusilcP, sc, contControl = cc,
+         design = "gender", fun = sim, k = 125)
```

The simulation results are inspected as usual. Note that the `aggregate()` method in this case returns values for each combination of contamination level and gender.

```
R> head(results)
```

```
  Run Sample Epsilon gender standard    Hill      PDC
1   1      1       1 0.0000   male 26.58067 26.50425 26.35969
2   1      1       1 0.0000 female 27.43355 27.03526 28.16992
3   2      1       1 0.0025   male 31.63593 29.23365 27.12430
4   2      1       1 0.0025 female 31.43540 27.77698 26.85896
5   3      1       1 0.0050   male 33.35950 31.07040 25.97415
6   3      1       1 0.0050 female 35.68710 34.03560 29.11359
```

```
R> aggregate(results)
```

```
   Epsilon gender standard    Hill      PDC
1   0.0000   male 25.94937 26.00709 25.85311
2   0.0025   male 30.44448 27.70155 26.01033
3   0.0050   male 33.54929 29.13202 26.16786
4   0.0075   male 36.76641 31.32342 26.49026
5   0.0100   male 39.42281 33.67944 26.53749
6   0.0000 female 27.30171 27.49442 27.41323
7   0.0025 female 31.68505 29.13643 27.61790
8   0.0050 female 35.49976 30.92128 27.91607
9   0.0075 female 38.51819 33.08778 28.09784
10  0.0100 female 41.47137 35.32935 27.97407
```

```
R> tv <- simSapply(eusilcP, "gender", function(x) gini(x$eqIncome)$value)
R> plot(results, true = tv)
```

If multiple contamination levels are used in a simulation study, the `plot()` method for the simulation results no longer produces box plots. Instead, the average results are plotted against the corresponding contamination levels, as shown in Figure 3.5. The plots show how the classical estimators move away from the references line as the contamination level increases, while the values obtained with the PDC estimator remain quite accurate.

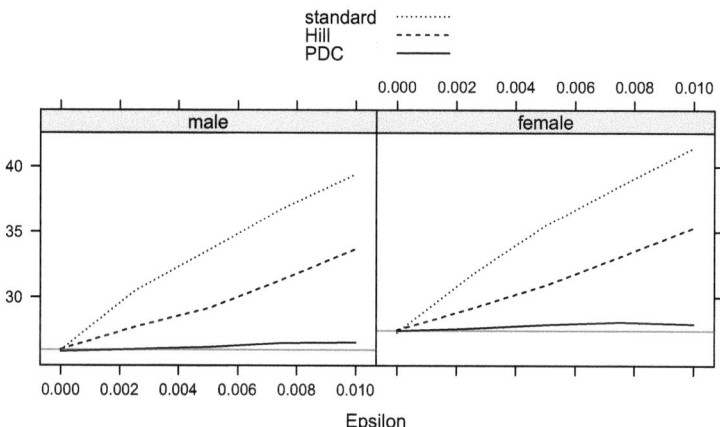

Figure 3.5: Simulation results for the simulation design with stratified sampling, multiple contamination levels and performing the simulations separately for each gender.

3.2.6 Inserting missing values

Survey data almost always contain a considerable amount of missing values. In close-to-reality simulation studies, the variability due to missing data therefore needs to be considered. Three types of missing data mechanisms are commonly distinguished in the literature (e.g., Little and Rubin 2002): missing completely at random (MCAR), missing at random (MAR) and missing not at random (MNAR). All three missing data mechanisms are implemented in the framework.

In the following example, missing values are inserted into the equivalized household income of non-contaminated households with MCAR, i.e., the households whose values are going to be set to NA are selected using simple random sampling. In order to compare the scenario without missing values to a scenario with missing values, the missing value rates 0% and 5% are used. In the latter case, the missing values are simply disregarded for fitting the Pareto distribution and estimating the Gini coefficient. Furthermore, the number of samples is reduced to 50 and only the contamination levels 0%, 0.5% and 1% are investigated to keep the computation time of this motivational example low.

With **simFrame**, only a control object for missing data needs to be defined and supplied to runSimulation(), the rest is done automatically by the framework. To apply these changes to a simulation study implemented with bespoke R code, yet another loop-like structure for the different missing value rates as well as changes in the data structure for the simulation results would be necessary.

3. Applications of statistical simulation in the case of EU-SILC

```
R> set.seed(12345)
R> sc <- SampleControl(design = "region", grouping = "hid",
+      size = c(75, 250, 250, 125, 200, 225, 125, 150, 100),
+      k = 50)
R> cc <- DCARContControl(target = "eqIncome", epsilon = c(0,
+      0.005, 0.01), dots = list(mean = 5e+05, sd = 10000))
R> nc <- NAControl(target = "eqIncome", NArate = c(0, 0.05))
R> results <- runSimulation(eusilcP, sc, contControl = cc,
+      NAControl = nc, design = "gender", fun = sim, k = 125)
```

As always, the `head()`, `aggregate()` and `plot()` methods are used to take a look at the simulation results. It should be noted that a column is added to the results that indicates the missing value rate used and that `aggregate()` in this example returns a value for each combination of contamination level, missing value rate and gender.

```
R> head(results)
```

	Run	Sample	Epsilon	NArate	gender	standard	Hill	PDC
1	1	1	0.000	0.00	male	26.58067	27.00998	26.26273
2	1	1	0.000	0.00	female	27.43355	27.92305	26.69034
3	2	1	0.000	0.05	male	26.62313	26.54198	26.01043
4	2	1	0.000	0.05	female	27.51209	26.83574	27.25464
5	3	1	0.005	0.00	male	33.71363	28.44824	26.46635
6	3	1	0.005	0.00	female	35.47508	28.48208	27.70783

```
R> aggregate(results)
```

	Epsilon	NArate	gender	standard	Hill	PDC
1	0.000	0.00	male	25.89948	25.99777	25.74944
2	0.005	0.00	male	33.52791	29.30477	26.14659
3	0.010	0.00	male	39.45422	32.74672	26.64929
4	0.000	0.05	male	25.88434	25.87824	25.80541
5	0.005	0.05	male	33.87975	29.60079	26.18759
6	0.010	0.05	male	39.99526	33.44462	26.31274
7	0.000	0.00	female	27.17769	27.30586	27.19275
8	0.005	0.00	female	35.46414	31.37099	27.98622
9	0.010	0.00	female	41.28625	35.22113	28.19677
10	0.000	0.05	female	27.16026	27.37710	27.20892
11	0.005	0.05	female	35.85305	31.56317	27.80455
12	0.010	0.05	female	41.86453	35.44025	27.98948

3.2 Application of different simulation designs to EU-SILC

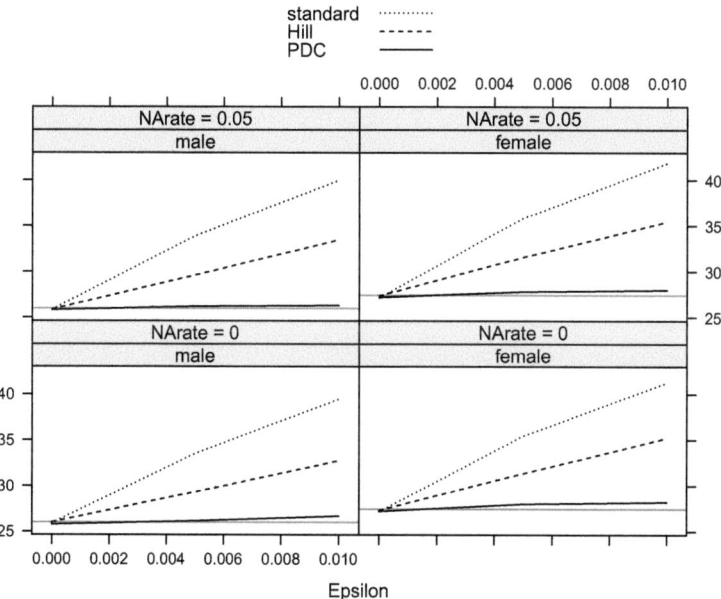

Figure 3.6: Simulation results for the simulation design with stratified sampling, multiple contamination levels, multiple missing value rates and performing the simulations separately for each gender.

```
R> tv <- simSapply(eusilcP, "gender", function(x) gini(x$eqIncome)$value)
R> plot(results, true = tv)
```

If multiple contamination levels and multiple missing value rates are used in the simulation study, conditional plots are produced by the plot() method for the simulation results. Figure 3.6 shows the resulting plots for this example. The bottom panels illustrate the scenario without missing values, while the scenario with 5% missing values is displayed in the top panels. In this case, there is not much of a difference in the results for the two scenarios.

3.2.7 Parallel computing

Statistical simulation is an *embarrassingly parallel* procedure, hence parallel computing can drastically reduce the computational costs. In **simFrame**, parallel computing is implemented using **snow** (Rossini et al. 2007, Tierney et al. 2008). Only minimal additional programming effort due to the use of **snow** is required to adapt the code from the previous example: to initialize the computer cluster, to ensure that all packages and objects

3. Applications of statistical simulation in the case of EU-SILC

are available on each worker process, to use the function `clusterRunSimulation()` instead of `runSimulation()` and to stop the computer cluster after the simulations. In addition, random number streams (e.g., L'Ecuyer et al. 2002, Sevcikova and Rossini 2009) should be used instead of the built-in random number generator.

```
R> cl <- makeCluster(4, type = "SOCK")
R> clusterEvalQ(cl, {
+     library("simFrame")
+     library("laeken")
+     data("eusilcP")
+ })
R> clusterSetupRNG(cl, seed = 12345)
R> sc <- SampleControl(design = "region", grouping = "hid",
+     size = c(75, 250, 250, 125, 200, 225, 125, 150, 100),
+     k = 50)
R> cc <- DCARContControl(target = "eqIncome", epsilon = c(0,
+     0.005, 0.01), dots = list(mean = 5e+05, sd = 10000))
R> nc <- NAControl(target = "eqIncome", NArate = c(0, 0.05))
R> clusterExport(cl, c("sc", "cc", "nc", "sim"))
R> results <- clusterRunSimulation(cl, eusilcP, sc, contControl = cc,
+     NAControl = nc, design = "gender", fun = sim, k = 125)
R> stopCluster(cl)
```

When the parallel computations are finished and the simulation results are obtained, they can be inspected as usual.

```
R> head(results)
```

	Run	Sample	Epsilon	NArate	gender	standard	Hill	PDC
1	1	1	0.000	0.00	male	26.20067	27.02017	23.66565
2	1	1	0.000	0.00	female	28.79194	29.23548	27.12933
3	2	1	0.000	0.05	male	26.19328	24.91570	24.07906
4	2	1	0.000	0.05	female	28.86860	27.38585	27.80012
5	3	1	0.005	0.00	male	34.46084	31.74470	24.87023
6	3	1	0.005	0.00	female	36.27429	32.14269	28.06137

```
R> aggregate(results)
```

	Epsilon	NArate	gender	standard	Hill	PDC
1	0.000	0.00	male	25.89996	25.98977	25.86451
2	0.005	0.00	male	33.56743	29.36361	26.39515

3.2 Application of different simulation designs to EU-SILC

```
3    0.010   0.00   male   39.40362 33.05926 26.68715
4    0.000   0.05   male   25.87909 25.86055 26.00109
5    0.005   0.05   male   33.94829 29.65456 26.32813
6    0.010   0.05   male   39.95535 33.24853 26.78947
7    0.000   0.00 female   27.38636 27.52210 27.48816
8    0.005   0.00 female   35.52688 31.30099 28.03385
9    0.010   0.00 female   41.35311 35.81549 28.67901
10   0.000   0.05 female   27.38459 27.51825 27.54063
11   0.005   0.05 female   35.87991 31.74678 28.18308
12   0.010   0.05 female   41.89804 36.21921 28.41367
```

```
R> tv <- simSapply(eusilcP, "gender", function(x) gini(x$eqIncome)$value)
R> plot(results, true = tv)
```

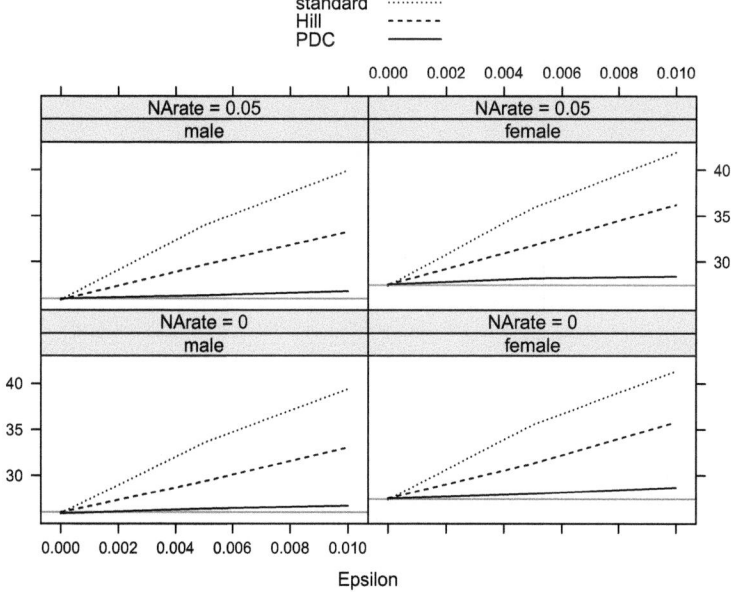

Figure 3.7: Simulation results obtained by parallel computing for the simulation design with stratified sampling, multiple contamination levels, multiple missing value rates and performing the simulations separately for each gender.

Figure 3.7 shows the simulation results obtained with parallel computing. The plots are, of course, very similar to the plots for the previous example in Figure 3.6, since the design of the simulation studies is the same.

3.3 Conclusions

In this paper, the use of the R package **simFrame** for different simulation designs has been demonstrated in a practical application. The full functionality of the framework for design-based simulation has been presented in various code examples. These examples showed that the framework allows researchers to make use of a wide range of simulation designs with only a few lines of code. In order to switch from one simulation design to another, only control objects need to be defined or modified. Even moving from basic to highly complex designs therefore requires only minimal changes to the code. With bespoke R code, such modifications would often need a considerable amount of programming. Furthermore, parallel computing with **simFrame** can easily be done based on package **snow**.

Besides the functionality for carrying out simulation studies, methods for several frequently used generic functions are available for inspecting or summarizing the simulation results. Most notably, a suitable plot method of the simulation results is selected automatically depending on their structure.

Due to this flexibility, **simFrame** is widely applicable for gaining insight into the quality of statistical methods and is a valuable addition to a researcher's toolbox.

Computational details All computations in this paper were performed using **Sweave** (Leisch 2002a,b) with R version 2.12.0 and **simFrame** version 0.3.6. The most recent version of the package is always available from CRAN (the Comprehensive R Archive Network, http://cran.R-project.org), and (an up-to-date version of) this paper is also included as a package vignette (Leisch 2003).

Acknowledgments This work was partly funded by the European Union (represented by the European Commission) within the 7th framework programme for research (Theme 8, Socio-Economic Sciences and Humanities, Project AMELI (Advanced Methodology for European Laeken Indicators), Grant Agreement No. 217322). Visit http://ameli.surveystatistics.net for more information on the project.

Chapter 4

Contamination models in the R package simFrame for statistical simulation

Published in *Computer Data Analysis and Modeling: Complex Stochastic Data and Systems* (Alfons et al. 2010d).

Andreas Alfons[a], Matthias Templ[a,b], Peter Filzmoser[a]

[a] Department of Statistics and Probability Theory, Vienna University of Technology
[b] Methods Unit, Statistics Austria

Abstract Due to the complexity of robust statistical methods, simulation studies are widely used to gain insight into the quality of these procedures. The R package **simFrame** is an object-oriented framework for statistical simulation with special emphasis on applications in robust statistics. Contamination is thereby modeled as a two-step process. Furthermore, the existing framework may be extended with user-defined contamination models.

4.1 Introduction

Robust statistical methods are becoming increasingly complex, therefore obtaining analytical results about their properties is becoming more and more time-consuming and difficult—often virtually impossible. On the other hand, computers are ever getting faster and cheaper. Therefore simulation studies are widely used by researchers to gain insight into the quality of the developed methods in different situations.

Two main concepts for simulation studies are distinguished in the literature: *model-based* and *design-based* simulation. In model-based simulation, data are generated repeatedly from a certain distribution. In every iteration, different methods are applied and quantities of interest are computed for comparison. Reference values can be obtained from the underlying theoretical distribution where appropriate. In design-based simulation, samples are drawn repeatedly from a finite population. Since real population data are only in few cases available to researchers, synthetic populations need to be generated (Alfons et al. 2010b). In every iteration, certain estimators such as indicators are computed. Where appropriate, these can be compared to the true population values.

When investigating robust methods, outliers need to be included. For model-based simulation, reference values are then computed from the theoretical distribution of the non-contaminated data. For design-based simulation, the situation is more complex (Alfons et al. 2009). The most realistic scenario would be to include outliers in the population data. However, total control over the amount of contamination is required for proper evaluation of robust methods. It is therefore suggested to generate outliers in the samples. In any case, reference values are computed from the non-contaminated population values.

The R package **simFrame** (Alfons et al. 2010c) is a general framework for simulation studies in statistics. Its object-oriented implementation provides clear interfaces for extensions by the user. One of the main advantages of **simFrame** is that simulation studies can be defined in terms of *control objects*. For large research projects, this ensures that results obtained by different partners are comparable.

4.2 Contamination models in simFrame

In the literature on robust statistics, the distribution F of contaminated data is typically modeled as a mixture of distributions

$$F = (1-\varepsilon)G + \varepsilon H, \tag{4.1}$$

where ε denotes the *contamination level*, G is the distribution of the non-contaminated part of the data and H is the distribution of the contamination (Maronna et al. 2006). As a consequence, outliers may be modeled by a two-step process (Hulliger and Schoch 2009b). The first step is to select observations to be contaminated, the second is to model the distribution of the outliers. Let n be the number of observations, p the number of variables, and let $\boldsymbol{x}_i = (x_{i1}, \ldots, x_{ip})$, $i = 1, \ldots, n$, denote the observations.

1. Let O_i, $i = 1, \ldots, n$, be an indicator whether an observation is an outlier ($O_i = 1$) or not ($O_i = 0$). The situation that the probability distribution of O_i does not depend on any other variables, i.e., that

$$P(O_i = 1 | x_{i1}, \ldots, x_{ip}) = P(O_i = 1), \qquad i = 1, \ldots, n \tag{4.2}$$

may be called *outlying completely at random* (OCAR). If Equation (4.2) is violated, i.e., if the probability distribution of O_i depends on observed information, the situation may be called *outlying at random* (OAR).

2. Let $I_c := \{i = 1, \ldots, n : O_i = 1\}$ be the index set of the observations to be contaminated, and let $\boldsymbol{x}_i^* = (x_{i1}^*, \ldots, x_{ip}^*)$ denote the true (i.e., non-contaminated) values of $\boldsymbol{x}_i$, $i \in I_c$. If the distribution H does not depend on the true values, i.e., if $\boldsymbol{x}_i \sim H(x_1, \ldots, x_p)$, $i \in I_c$, the outliers may be called *contaminated completely at random* (CCAR). On the other hand, if H depends on the true values, i.e., if $\boldsymbol{x}_i \sim H(x_1, \ldots, x_p, x_{i1}^*, \ldots, x_{ip}^*)$, $i \in I_c$, the outliers may be called *contaminated at random* (CAR).

The package **simFrame** is implemented in the open-source statistical environment and programming language R (R Development Core Team 2010). Taking advantage of object-oriented programming, the control classes `DCARContControl` and `DARContControl` determine how contamination is handled in simulation studies (see the example in Section 4.3). `DCARContControl` may be used for OCAR-CCAR and OAR-CCAR models, whereas `DARContControl` corresponds to OCAR-CAR and OAR-CAR. Additional contamination models may be added in the future. However, the object-oriented design further allows contamination models to be implemented by the user. The programming interfaces for such extensions are described in detail in Alfons et al. (2010c).

4.3 Example: Outlier detection

This simple motivational example for the usage of **simFrame** is a comparison of outlier detection using classical and robust estimation of location and scatter. The robust estimates are obtained with the fast MCD (Rousseeuw and Van Driessen 1999) implementation in package **rrcov** (Todorov and Filzmoser 2009).

Data are generated in each of the 100 simulation runs from a two-dimensional normal distribution. Varying the contamination level between 10% and 30% in steps of 5%, the contaminated data are generated from a normal distribution with a shifted mean (OCAR-CCAR). In the function to be executed in every iteration, the percentages of false negatives and false positives are computed. Note that the default tuning parameter 0.5 is used for the MCD. For reproducibility of the simulation results, the seed of the random number generator is set before running the simulation study.

```
R> sigma <- matrix(c(1, 0.5, 0.5, 1), 2, 2)
R> dc <- DataControl(size = 100, distribution = rmvnorm,
+        dots = list(sigma = sigma))
```

4. Contamination models in the R package simFrame

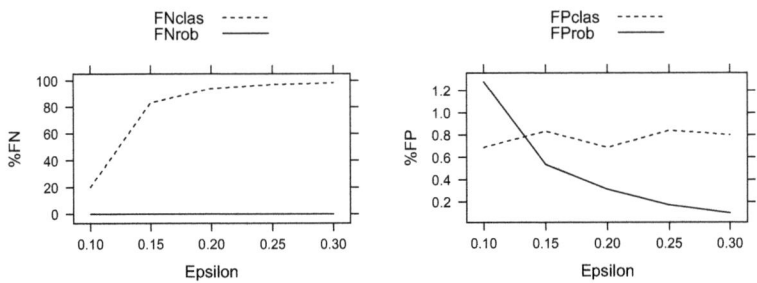

Figure 4.1: Average proportions of false negatives (*left*) and false positives (*right*).

```
R> cc <- DCARContControl(epsilon = seq(0.1, 0.3, by = 0.05),
+       distribution = rmvnorm, dots = list(mean = c(5, -5),
+           sigma = sigma))
R> sim <- function(x, q) {
+       clas <- Cov(x[, 1:2])
+       rob <- CovMcd(x[, 1:2])
+       dclas <- mahalanobis(x[, 1:2], clas@center, clas@cov)
+       drob <- mahalanobis(x[, 1:2], rob@center, rob@cov)
+       outclas <- dclas > q
+       outrob <- drob > q
+       nout <- length(which(x$.contaminated))
+       ngood <- nrow(x) - nout
+       c(FNclas = length(which(!outclas & x$.contaminated))/nout,
+           FNrob = length(which(!outrob & x$.contaminated))/nout,
+           FPclas = length(which(outclas & !x$.contaminated))/ngood,
+           FProb = length(which(outrob & !x$.contaminated))/ngood) *
+           100
+ }
R> set.seed(12345)
R> result <- runSimulation(dc, nrep = 100, contControl = cc,
+       fun = sim, q = qchisq(0.975, df = 2))
R> plot(result, select = c("FNclas", "FNrob"), ylab = "%FN")
R> plot(result, select = c("FPclas", "FProb"), ylab = "%FP")
```

In **simFrame**, a suitable graphical representation of the results is selected automatically depending on their structure. Figure 4.1 shows plots of the average proportions of false negatives (*left*) and false positives (*right*). The plots, of course, clearly favor the MCD over classical estimation.

4.4 Conclusions

The package **simFrame** is an object-oriented framework for simulation studies in the statistical environment R. Different contamination models are implemented using control classes. The flexible framework further allows additional contamination models to be implemented by the user. Hence **simFrame** is widely applicable in the field of robust statistics.

Acknowledgments This work was partly funded by the European Union within the 7th framework programme for research (Project AMELI, Grant Agreement No. 217322).

4. Contamination models in the R package **simFrame**

Chapter 5

Simulation of close-to-reality population data for household surveys with application to EU-SILC[1]

Revision submitted to the journal *Statistical Methods & Applications*.

Andreas Alfons[a], Stefan Kraft[a,b], Matthias Templ[a,c], Peter Filzmoser[a]

[a] Department of Statistics and Probability Theory, Vienna University of Technology
[b] now at the Institute for Quantitative Asset Management
[c] Methods Unit, Statistics Austria

Abstract Statistical simulation in survey statistics is usually based on repeatedly drawing samples from population data. Furthermore, population data may be used in courses on survey statistics to explain issues regarding, e.g., sampling designs. Since the availability of real population data is in general very limited, it is necessary to generate synthetic data for such applications. The simulated data need to be as realistic as possible, while at the same time ensuring data confidentiality. This paper proposes a method for generating close-to-reality population data for complex household surveys. The procedure consists of four steps for setting up the household structure, simulating categorical variables, simulating continuous variables and splitting continuous variables into different components. It is not required to perform all four steps so that the framework is applicable to a broad class of surveys. In addition, the proposed method is evaluated in an application to the European Union Statistics on Income and Living Conditions (EU-SILC).

---

[1]This work was partly funded by the European Union (represented by the European Commission) within the 7th framework programme for research (Theme 8, Socio-Economic Sciences and Humanities, Project AMELI (Advanced Methodology for European Laeken Indicators), Grant Agreement No. 217322). Visit http://ameli.surveystatistics.net for more information on the project.

5. Simulation of close-to-reality population data

Keywords Synthetic data, Simulation, Survey statistics, EU-SILC

5.1 Introduction

Survey data contain variability due to sampling, imputation of missing values, measurement errors and editing. Statistical simulation in survey statistics therefore often follows a *close-to-reality* approach (see, e.g., Münnich et al. 2003b), i.e., the behavior of the developed methodology for a specific survey is investigated by repeatedly drawing samples from population data with the sampling method and weighting scheme used in practice. Population data may thus form the basis for a realistic framework to compare statistical methods under different settings. In particular, the estimation of indicators needed for policy decisions may be investigated with respect to different sampling designs or common data problems such as measurement errors or missing values.

In teaching, population data may support courses on topics such as sampling, statistical modeling or indicator estimation. Again, real-world situations could be considered by drawing samples from close-to-reality populations. Issues regarding, e.g., the sampling design or inhomogeneities in the data can be explained using real-world applications.

However, real population data are typically limited to census or register data. Only in exceptions are suitable population data available to researchers. The remedy of this problem is to generate synthetic populations from existing survey data.

Simulation of population microdata is closely related to the field of *microsimulation* (e.g., Clarke 1996), which is a well-established methodology within the social sciences, although the aims are quite different. Microsimulation models attempt to reproduce the behavior of individual units such as persons, households or firms over the course of many years for policy analysis purposes. Hence they are highly complex and time-consuming. Survey statisticians, on the other hand, need synthetic populations as a basis for extensive simulation studies on the behavior of their statistical methods. Fast computation is thus favored to over-complex models.

An alternative approach for the generation of synthetic data sets is discussed by Rubin (1993). He addresses the confidentiality problem connected with the release of publicly available microdata and proposes the generation of fully synthetic microdata sets using multiple imputation. Raghunathan et al. (2003), Drechsler et al. (2008) and Reiter (2009) discuss this approach in more detail. However, their approach does not allow to generate categories that are not represented in the original sample, nor do they investigate the possible generation of structural zeros in combinations of variables. Moreover, some basic variables from the real population data are required as auxiliary information.

The generation of population microdata for selected surveys as a basis for Monte Carlo simulation is described by Münnich et al. (2003b) and Münnich and Schürle (2003). Nevertheless, their framework was developed for household surveys with large sample sizes that contain mainly categorical variables. All steps of the procedure are performed separately for each stratum of the sampling design. The household structure is thereby simulated in two steps. First, the household sizes are drawn from the observed conditional distributions within the strata. Second, the age and gender structure of the population households is generated by resampling households of the same size from the respective strata in the sample. Additional categorical variables are then simulated by random draws from the observed conditional distributions of their multivariate realizations within each combination of stratum, age (or age category) and gender. Also continuous variables are modeled separately for each combination of stratum and outcomes from certain influential variables.

In any case, this framework has been modified and extended in order to be applicable to more complex surveys such as the well-known *European Union Statistics on Income and Living Conditions* (EU-SILC). Please note that while it would be interesting to establish a theoretical relationship between the goodness of the statistical models and the resulting populations, such an analysis is out of scope for this paper due to the large number of models involved. Instead, the proposed procedure is evaluated by means of simulation.

The rest of the paper is organized as follows. In Section 5.2, the proposed data simulation method is described in great detail. Diagnostic plots and results from extensive simulation studies in an application to EU-SILC are presented in Section 5.3. The final Section 5.4 concludes.

5.2 Simulation of synthetic populations

The data simulation method proposed in this paper is motivated by the *European Union Statistics on Income and Living Conditions* (EU-SILC; see Section 5.3), but since it is designed to manage all difficulties of this highly complex survey, it is also applicable to many other household surveys. In any case, the following conditions need to be respected when simulating population data (Münnich et al. 2003b, Münnich and Schürle 2003):

- Actual sizes of regions and strata need to be reflected.

- Marginal distributions and interactions between variables should be represented correctly.

- Heterogeneities between subgroups, especially regional aspects, should be allowed.

- Pure replication of units from the underlying sample should be avoided, as this generally leads to extremely small variability of units within smaller subgroups.

- Data confidentiality must be ensured.

5. Simulation of close-to-reality population data

In the case of EU-SILC, another problem needs to be considered. As the name suggests, EU-SILC contains information about income, which is split into different income components. The data simulation method must thus ensure that such a breakdown of variables is done in a realistic manner.

Since some of the above conditions are conflicting with one another, generating completely realistic populations seems an impossible task. Nevertheless, being as close to reality as possible suffices for drawing meaningful conclusions from simulation studies.

Our procedure is based on the ideas of Münnich et al. (2003b) and Münnich and Schürle (2003). However, they mainly consider the generation of categorical variables for specific surveys such as the German Microcensus, with only a few simple extensions to continuous variables. The proposed method uses modifications of their framework and both improves and extends the simulation scheme such that it can be applied to a much broader class of household surveys. This in particular includes surveys with relatively small sample sizes or with complex continuous variables or components thereof. In general, the procedure consists of four steps:

1. Setup of the household structure

2. Simulation of categorical variables

3. Simulation of continuous variables

4. Splitting continuous variables into components

While the propositions of Münnich et al. (2003b) and Münnich and Schürle (2003) are only slightly modified in Step 1, an entirely different approach is used in Steps 2 and 3. In addition, Step 4 constitutes a new development motivated by EU-SILC. Having different stages provides maximum flexibility of the framework. Depending on the specific survey, not all four steps need to be carried out.

It is important to note that the proposed data generation method relies solely on the underlying sample data, no auxiliary information (e.g., available census data) is required. Stratification allows to account for heterogeneities such as regional differences. Furthermore, sample weights are considered in each step to ensure high similarity of expected and realized values. Concerning data confidentiality, a detailed analysis of the framework using different worst case scenarios is carried out in Templ and Alfons (2010). The conclusion of this analysis is that the synthetic population data are confidential and may be distributed to the public.

In the following sections, the different steps of the procedure are described in detail. Section 5.2.5 then briefly discusses the implementation of the procedure in R (R Development Core Team 2010).

5.2.1 Setup of the household structure

The household structure is simulated separately for each combination of stratum k and household size l. First, the number of households M_{kl} is estimated using the Horvitz-Thompson estimator (Horvitz and Thompson 1952):

$$\hat{M}_{kl} := \sum_{h \in H_{kl}^S} w_h, \quad (5.1)$$

where H_{kl}^S denotes the index set of households in stratum k of the survey data with household size l, and w_h, $h \in H_{kl}^S$, are the corresponding household weights. Similarly, let H_{kl}^U be the respective index set of households in the population data such that $|H_{kl}^U| = \hat{M}_{kl}$. To prevent unrealistic structures in the population households, basic information from the survey households is resampled. Let x_{hij}^S and x_{hij}^U denote the value of person i from household h in variable j for the sample and population data, respectively, and let the first p_1 variables contain the basic information on the household structure. For each population household $h \in H_{kl}^U$, a survey household $h' \in H_{kl}^S$ is selected with probability $w_{h'}/\hat{M}_{kl}$ and the household structure is set to

$$x_{hij}^U := x_{h'ij}^S, \quad i = 1, \ldots, l, \ j = 1, \ldots, p_1. \quad (5.2)$$

Alias sampling (Walker 1977) is well suited for our purpose, as it is very fast for a large number of sampled elements. Furthermore, as few variables as possible should be adopted by the persons in the resampled households for disclosure reasons. Our suggestion is to use only age and gender information, which is typically available in household surveys.

5.2.2 Simulation of categorical variables

For simulating additional categorical variables, the approach by Münnich et al. (2003b) and Münnich and Schürle (2003) is based on estimating conditional distributions directly by the corresponding relative frequency distributions in the underlying sample. It therefore requires a rather large sample size and is not very flexible (see Section 5.3.2; cf. Kraft 2009). In particular, it does not allow to generate combinations that do not occur in the sample. To overcome these shortcomings, the proposed approach estimates conditional distributions with multinomial logistic regression models.

Let $\boldsymbol{x}_j^S = (x_{1j}^S, \ldots, x_{nj}^S)'$ and $\boldsymbol{x}_j^U = (x_{1j}^U, \ldots, x_{Nj}^U)'$ denote the variables in the sample and population, respectively, where n and N give the corresponding number of individuals. The additional categorical variables are thereby given by the indices $p_1 < j \leq p_2$. Furthermore, the personal sample weights are denoted by $\boldsymbol{w} = (w_1, \ldots, w_n)'$. Multinomial logistic regression models are fitted for each stratum separately. Due to limited space, a detailed

5. Simulation of close-to-reality population data

mathematical description of these models cannot be provided in this paper, but can be found in, e.g., Simonoff (2003).

The following procedure is performed for each stratum k and each variable to be simulated, given by the index j, $p_1 < j \leq p_2$. Let I_k^S and I_k^U be the index sets of individuals in stratum k for the survey and population data, respectively. The survey data given by the indices in I_k^S is used to fit the model with response $\boldsymbol{x}_j^S$ and predictors $\boldsymbol{x}_1^S, \ldots, \boldsymbol{x}_{j-1}^S$, thereby considering the sample weights w_i, $i \in I_k^S$. Furthermore, let $\{1, \ldots, R\}$ be the set of possible outcome categories of the response variable. In particular, the number of possible outcomes is denoted by R. For every individual $i \in I_k^U$, the conditional probabilities $p_{ir}^U := P(x_{ij}^U = r | x_{i1}^U, \ldots, x_{i,j-1}^U)$ are estimated by

$$
\begin{aligned}
\hat{p}_{i1}^U &:= \frac{1}{1 + \sum_{r=2}^{R} \exp(\hat{\beta}_{0r} + \hat{\beta}_{1r} x_{i1}^U + \ldots + \hat{\beta}_{j-1,r} x_{i,j-1}^U)}, \\
\hat{p}_{ir}^U &:= \frac{\exp(\hat{\beta}_{0r} + \hat{\beta}_{1r} x_{i1}^U + \ldots + \hat{\beta}_{j-1,r} x_{i,j-1}^U)}{1 + \sum_{r=2}^{R} \exp(\hat{\beta}_{0r} + \hat{\beta}_{1r} x_{i1}^U + \ldots + \hat{\beta}_{j-1,r} x_{i,j-1}^U)}, \quad r = 2, \ldots, R,
\end{aligned}
\quad (5.3)
$$

where $\hat{\beta}_{0r}, \ldots, \hat{\beta}_{j-1,r}$, $r = 2, \ldots, R$, are the estimated coefficients (see, e.g., Simonoff 2003). The values of $\boldsymbol{x}_j^U$ for the individuals $i \in I_k^U$ are then drawn from the corresponding conditional distributions.

Note that for simulating the jth variable, $p_1 < j \leq p_2$, the $j-1$ previous variables are used as predictors. This means that the order of the additional categorical variables may be relevant. However, once such a variable is generated in the population, that information should certainly be used for simulating the remaining variables. In our application to EU-SILC, changing the order of the variables did not produce significantly different results (not shown). Alternatively, the procedure could be continued iteratively once all additional variables are available in the population, in each step using all other variables as predictors. Nevertheless, such a procedure would be computationally very expensive for real-life sized population data.

Estimating the conditional distributions with multinomial logistic regression models allows to simulate combinations that do not occur in the sample but are likely to occur in the true population. Such combinations are called *random zeros*, as opposed to *structural zeros*, which are impossible to occur (e.g., Simonoff 2003). For close-to-reality populations, such structural zeros need to be reflected. This can be done by setting $p_{ir'}^U := 0$, where r' is an impossible value for x_{ij} given $x_{i1}, \ldots, x_{i,j-1}$, and adjusting the other probabilities so that $\sum_{r=1}^{R} p_{ir}^U = 1$.

5.2.3 Simulation of continuous variables

Continuing the notation from the previous section, let x_j^S and x_j^U, $p_2 < j \leq p_3$, denote the continuous variables. Two different approaches are presented in the following. Both are able to handle semi-continuous variables, i.e., variables that contain a large amount of zeros.

Multinomial model with random draws from resulting categories

This approach is based on the simulation of categorical variables described in the previous section. The following steps are performed for each variable to be simulated, given by the index j, $p_2 < j \leq p_3$. First, the variable x_j^S is discretized. This is done in a different manner for continuous and semi-continuous variables. For continuous variables, $R+1$ breakpoints $b_1 < \ldots < b_{R+1}$ are used to define the discretized variable $\boldsymbol{y}^S = (y_1^S, \ldots, y_n^S)'$ as

$$y_i^S := \begin{cases} 1 & \text{if } b_1 \leq x_{ij}^S \leq b_2, \\ r & \text{if } b_r < x_{ij}^S \leq b_{r+1},\ r = 2, \ldots, R. \end{cases} \quad (5.4)$$

For semi-continuous variables, zero is a category of its own, and breakpoints for negative and positive values are distinguished. Let $b_{R^-+1}^- < \ldots < b_1^- = 0 = b_1^+ < \ldots < b_{R^++1}^+$ be the breakpoints. Then $\boldsymbol{y}^S$ is defined as

$$y_i^S := \begin{cases} -r & \text{if } R^- > 0 \text{ and } b_{r+1}^- \leq x_{ij}^S < b_r^-,\ r = R^-, \ldots, 1, \\ 0 & \text{if } x_{ij}^S = 0, \\ r & \text{if } R^+ > 0 \text{ and } b_r^+ < x_{ij}^S \leq b_{r+1}^+,\ r = 1, \ldots, R^+. \end{cases} \quad (5.5)$$

Note that the cases of only non-negative or non-positive values in x_j^S are considered in (5.5).

Multinomial logistic regression models with response $\boldsymbol{y}^S$ and predictors $\boldsymbol{x}_1^S, \ldots, \boldsymbol{x}_{j-1}^S$ are then fitted for every stratum k separately, as described in the previous section, in order to simulate the values of the categorized population variable $\boldsymbol{y}^U = (y_1^U, \ldots, y_N^U)'$.

Finally, the values of x_j^U are generated by random draws from uniform distributions within the corresponding categories of $\boldsymbol{y}^U$. For continuous variables, the values of individual $i = 1, \ldots, N$ are generated as

$$x_{ij}^U \sim U(b_r, b_{r+1}) \text{ if } y_i^U = r. \quad (5.6)$$

For semi-continuous variables, the values of individual $i = 1, \ldots, N$ are set to $x_{ij}^U := 0$ if $y_i^U = 0$, while the non-zero observations are generated as

$$x_{ij}^U \sim \begin{cases} U(b_{r+1}^-, b_r^-) & \text{if } y_i^U = -r < 0, \\ U(b_r^+, b_{r+1}^+) & \text{if } y_i^U = r > 0. \end{cases} \quad (5.7)$$

5. Simulation of close-to-reality population data

The idea behind this approach is to divide the data into relatively small subsets. If the intervals are too large, using uniform distributions may be an oversimplification. However, the advantage of this approach is that it allows the breakpoints for the discretization to be chosen in such a way that the empirical distribution is well reflected in the simulated population variable. It thereby needs to be considered that the larger the number of breakpoints, the higher the computation time. Quantiles in steps of 10% are reasonable default values for the breakpoints, while the fit in the tails of the distribution may be improved by also using the 1%, 5%, 95% and 99% quantiles. Note that sufficient accuracy in some applications may already be reached with larger steps in the middle part of the distribution (see Section 5.3).

When simulating variables that contain extreme values, such as income, *tail modeling* should be considered. In that case, values from the largest categories could be drawn from a generalized Pareto distribution (GPD). The cumulative distribution function of the GPD is defined as

$$F_{\mu,\sigma,\xi}(x) = \begin{cases} 1 - \left(1 + \dfrac{\xi(x-\mu)}{\sigma}\right)^{-\frac{1}{\xi}}, & \xi \neq 0, \\ 1 - \exp\left(-\dfrac{x-\mu}{\sigma}\right), & \xi = 0, \end{cases}$$

where μ is the location parameter, $\sigma > 0$ is the scale parameter and ξ is the shape parameter. The range of x is $x \geq 0$ when $\xi \geq 0$ and $\mu \leq x \leq \mu - \frac{\sigma}{\xi}$ when $\xi < 0$. See, e.g., Embrechts et al. (1997) for details on the *peaks over threshold* approach for fitting the GPD. Note that other distributions may be used for tail modeling as well (see, e.g., Kleiber and Kotz 2003). Nevertheless, if the purpose of such a population is comparing different estimators in a simulation study, it is important to note that using a GPD for the tails favors estimators that incorporate generalized Pareto tail modeling over other types of estimators.

(Two-step) regression model with random error terms

The second approach is based on linear regression combined with random error terms. Semi-continuous variables are thereby simulated using a two-step model. The following procedure is repeated for each variable to be simulated, given by the index j, $p_2 < j \leq p_3$.

For semi-continuous variables, the first step is to simulate whether x_{ij}^U, $i = 1, \ldots, N$, is zero or not. This is done by fitting logistic regression models (see, e.g., Simonoff 2003) for each stratum separately. The binary response variable $\boldsymbol{y}^S = (y_1^S, \ldots, y_n^S)'$ is defined as

$$y_i^S := \begin{cases} 0 & \text{if } x_{ij} = 0, \\ 1 & \text{else.} \end{cases} \tag{5.8}$$

For each stratum k, the observations given by the index set I_k^S are used to fit the model with response $\boldsymbol{y}^S$ and predictors $\boldsymbol{x}_1^S, \ldots, \boldsymbol{x}_{j-1}^S$. The sample weights w_i, $i \in I_k^S$, are considered in the model fitting process by using a weighted maximum likelihood approach. For every

5.2 Simulation of synthetic populations

individual $i \in I_k^U$, the conditional probabilities $p_i^U := P(y_i^U = 1 | x_{i1}^U, \ldots, x_{i,j-1}^U)$ that x_{ij}^U is non-zero are estimated by

$$\hat{p}_i^U := \frac{\exp(\hat{\beta}_0 + \hat{\beta}_1 x_{i1}^U + \ldots + \hat{\beta}_{j-1} x_{i,j-1}^U)}{1 + \exp(\hat{\beta}_0 + \hat{\beta}_1 x_{i1}^U + \ldots + \hat{\beta}_{j-1} x_{i,j-1}^U)}, \qquad (5.9)$$

where $\hat{\beta}_0, \ldots, \hat{\beta}_{j-1}$ are the estimated coefficients (e.g., Simonoff 2003). The values y_i^U, $i \in I_k^U$, are then drawn from the corresponding conditional distributions. Consequently, the zeros in the simulated semi-continuous variable are given by $x_{ij}^U := 0$ if $y_i^U = 0$. For the second step, the non-zero observations are indicated by $\tilde{I}_k^S := \{i \in I_k^S : y_i^S = 1\}$ and $\tilde{I}_k^U := \{i \in I_k^U : y_i^U = 1\}$.

For continuous variables, on the other hand, $\tilde{I}_k^S := I_k^S$ and $\tilde{I}_k^U := I_k^U$ are used in the following. Linear regression models are fitted for every stratum separately. In order to obtain more robust models, trimming parameters α_1 and α_2 are introduced. The following procedure is carried out for each stratum k. Let the observations to be used for fitting the model be given by the index set $I_{\alpha_1}^{\alpha_2} := \{i \in \tilde{I}_k^S : q_{\alpha_1} < x_{ij} < q_{1-\alpha_2}\}$, where q_{α_1} and $q_{1-\alpha_2}$ are the corresponding α_1 and $1 - \alpha_2$ quantiles, respectively. The linear model is then given by

$$x_{ij}^S = \beta_0 + \beta_1 x_{i1}^S + \ldots + \beta_{j-1} x_{i,j-1}^S + \varepsilon_i^S, \quad i \in I_{\alpha_1}^{\alpha_2}, \qquad (5.10)$$

where ε_i^S are random error terms. Using the weighted least squares approach with weights w_i, $i \in I_{\alpha_1}^{\alpha_2}$, coefficients $\hat{\beta}_0, \ldots, \hat{\beta}_{j-1}$ are obtained (see, e.g., Weisberg 2005) and the population values are estimated by

$$\hat{x}_{ij}^U = \hat{\beta}_0 + \hat{\beta}_1 x_{i1}^U + \ldots + \hat{\beta}_{j-1} x_{i,j-1}^U + \varepsilon_i^U, \quad i \in \tilde{I}_k^U. \qquad (5.11)$$

The random error terms ε_i^U need to be added since otherwise individuals with the same set of predictor values would receive the same value in x_{ij}^U. There are two suggestions on how to generate the random error terms:

- Use random draws from the residuals

$$\hat{r}_i^S = x_{ij}^S - \left(\hat{\beta}_0 + \hat{\beta}_1 x_{i1}^S + \ldots + \hat{\beta}_{j-1} x_{i,j-1}^S \right), \quad i \in I_{\alpha_1}^{\alpha_2}. \qquad (5.12)$$

- Use random draws from a normal distribution $\mathcal{N}(\mu, \sigma^2)$. The parameters μ and σ are thereby estimated robustly with median and MAD, respectively.

The first approach is more data-driven, while the second approach is in accordance with the theoretical assumption of normally distributed errors. For both, the trimming parameters α_1 and α_2 need to be selected carefully. If they are too small, very large random error terms due to outliers may result in large deviations especially in the tails of the distribution. If they are too large, the random error terms may not introduce enough variability. In the application to EU-SILC, $\alpha_1 = \alpha_2 = 0.01$ appeared to be a reasonable choice.

For variables such as income, a log-transformation may be beneficial before fitting the linear model. Equation (5.10) is then changed to

$$\log x_{ij}^S = \beta_0 + \beta_1 x_{i1}^S + \ldots + \beta_{j-1} x_{i,j-1}^S + \varepsilon_i^S, \quad i \in I_{\alpha_1}^{\alpha_2}. \tag{5.13}$$

In that case, the population values are estimated by

$$\hat{x}_{ij}^U = \exp(\hat{\beta}_0 + \hat{\beta}_1 x_{i1}^U + \ldots + \hat{\beta}_{j-1} x_{i,j-1}^U + \varepsilon_i^U), \quad i \in \tilde{I}_k^U. \tag{5.14}$$

However, the log-transformation causes problems with negative values, which is realistic for income (losses from self employment, see the example with EU-SILC data in Section 5.3). A simple remedy is of course to add a constant $c > 0$ to x_{ij}^S to obtain positive values, i.e., to use $\log(x_{ij}^S + c)$ in the left-hand side of Equation (5.13). This constant then needs to be subtracted from the right hand side of Equation (5.14). Another possibility is to combine the two presented approaches for simulating (semi-)continuous variables. A multinomial model with one category for positive values and certain categories for non-positive values is applied in the first step. Positive values are then simulated using a linear model for the log-transformed data, while negative values are drawn from uniform distributions within the respective simulated categories.

5.2.4 Splitting continuous variables into components

The procedure for simulating components of continuous variables is motivated by EU-SILC data, which contain information on various income components. When simulating components, the following problems need to be considered (cf. Kraft 2009). Even for a moderate number of components, it may be too complex to consider all the dependencies between the components and the other variables, as well as between the components themselves. Moreover, sparseness of various components may be an issue, e.g., in EU-SILC data, most income components typically contain only few non-zero observations. To manage these problems, a simple but effective approach based on conditional resampling of fractions has been developed. Only very few highly influential categorical variables should thereby be considered for conditioning.

Let $\boldsymbol{z}^S = (z_1^S, \ldots, z_n^S)'$ and $\boldsymbol{z}^U = (z_1^U, \ldots, z_N^U)'$ denote the variable giving the total in the sample and population, respectively, and let $\boldsymbol{x}_j^S$ and $\boldsymbol{x}_j^U$, $p_3 < j \leq p_4$, denote the variables containing the components. First, the fractions of the components with respect to the total

5.2 Simulation of synthetic populations

are computed for the sample:

$$y_{ij}^S := \frac{x_{i,p_3+j}^S}{z_i^S}, \quad i \in I_r^S, \; j = 1, \ldots, p_4 - p_3. \tag{5.15}$$

For the second step, let J_c be the index set of the conditioning variables. This step is performed separately for every combination of outcomes $\boldsymbol{r} = (r_j)_{j \in J_c}$. Let $I_r^S := \{i : x_{ij}^S = r_j \; \forall j \in J_c\}$ and $I_r^U := \{i : x_{ij}^U = r_j \; \forall j \in J_c\}$ be the index sets of individuals in the survey and population data, respectively, with the corresponding outcomes in the conditioning variables. For each individual $i \in I_r^U$ in the population, an individual $i' \in I_r^S$ from the survey data is selected with probability $w_{i'} / \sum_{i \in I_r^S} w_i$ and the values of the components are set to

$$x_{i,p_3+j}^U := z_i^U y_{i'j}^S, \quad j = 1, \ldots, p_4 - p_3. \tag{5.16}$$

If no observations for combination $\boldsymbol{r}$ exist in the sample, i.e., if $I_r^S = \emptyset$, a suitable donor $\boldsymbol{r}'$ is selected by minimizing a suitable distance measure such as the Manhattan distance $d_1(\boldsymbol{r}, \boldsymbol{s}) = \|\boldsymbol{r} - \boldsymbol{s}\|_1$. Then $I_r^S := I_{r'}^S$ is used in the above steps.

Resampling fractions has the advantage that it avoids unrealistic or unreasonable combinations in the simulated components. At the same time, it does not result in pure replication, as the absolute values for simulated individuals are in general quite different from the corresponding individuals in the underlying survey data.

5.2.5 Software

The proposed data simulation framework is implemented in the R package **simPopulation** (Alfons and Kraft 2010), which can be obtained from CRAN (the Comprehensive R Archive Network, http://CRAN.R-project.org). For maximum flexibility, the four steps of the procedure are available as separate functions. To generate populations for EU-SILC, a wrapper combining all four steps is implemented in order to provide a more convenient interface. Wrappers for other surveys can easily be defined by the user. In addition, functions to create diagnostic plots as shown in Section 5.3.1 are available. The latter are implemented using packages **vcd** (Meyer et al. 2006, 2010) and **lattice** (Sarkar 2008, 2010).

It would certainly be beneficial to present a line-by-line illustration of the R code for the application in Section 5.3. Nevertheless, the EU-SILC sample provided by Statistics Austria is confidential, thus the reader would not be able to reproduce the results. Furthermore, the additional explanation of the R code would render the length of the paper far from being reasonable. Therefore, detailed instructions for such an analysis and the generation of diagnostic plots are provided in a separate package vignette. If **simPopulation** is installed, the vignette can be viewed from within R with the following command:

```
R> vignette("simPopulation-eusilc")
```

Note that the vignette uses the synthetically generated example data from the package, hence the results presented there are reproducible.

5.3 Application to EU-SILC

The *European Union Statistics on Income and Living Conditions* (EU-SILC) is one of the most well-known panel surveys and is conducted in EU member states and other European countries. It is mainly used as data basis for the *Laeken indicators*, a set of indicators for measuring risk-of-poverty and social cohesion in European countries (cf. Atkinson et al. 2002).

The application of the proposed data simulation procedure to EU-SILC (limited to non-negative personal net income and income components) is described in more detail in Kraft (2009), where an extensive collection of results can be found as well. With the generalizations presented in this paper, however, it is also possible to simulate negative income or income components. The underlying survey data used in this section is the Austrian EU-SILC sample from 2006. Table 5.1 lists the variables to be included in the simulation and their possible outcomes. It should be noted that due to low frequencies of occurence, some categories of economic status and citizenship, respectively, have been combined. Such combined categories are marked with an asterisk (*) in Table 5.1. A complete description of variables in EU-SILC and possible outcomes can be found in Eurostat (2004).

Section 5.3.1 presents some diagnostic plots for comparing synthetic population data to the underlying sample. How well the characteristics of the original sample are reflected in such synthetic populations is further assessed by simulation in Section 5.3.2. These comparisons with the underlying sample are essential as this is the only real data available. In Section 5.3.3, the influence of different sample sizes and sampling designs on the proposed methodology is investigated by more extensive simulation studies.

5.3.1 Diagnostic plots for a single simulation

For setting up the household structure, households from the survey data are resampled conditional on region and household size. Sensible correlation structures within the households are ensured by resampling the variables age and gender, as recommended in Section 5.2.1. Afterwards, the variable age is categorized in order to use the resulting categories for the rest of the simulation. Variables that are categorized in the data simulation procedure are listed in Table 5.2, along with the respective categories. Besides age, the personal net income is discretized at a later stage. The age categories are chosen as a resonable tradeoff between accuracy and computation time (see also Section 5.3.2). Children below 16 are combined into one category since EU-SILC provides information for the remaining variables to be sim-

5.3 Application to EU-SILC

Table 5.1: Variables selected for the simulation of the Austrian EU-SILC population data.

Variable	Name	Possible outcomes	
Region	db040	1	Burgenland
		2	Lower Austria
		3	Vienna
		4	Carinthia
		5	Styria
		6	Upper Austria
		7	Salzburg
		8	Tyrol
		9	Vorarlberg
Household size	hsize		Number of persons in household
Age	age		Age (for the previous year) in years
Gender	rb090	1	Male
		2	Female
Self-defined current economic status	pl030	1	Working full-time
		2	Working part-time
		3	Unemployed
		4	Pupil, student, further training or unpaid work experience or in compulsory military or community service*
		5	In retirement or in early retirement or has given up business
		6	Permanently disabled or/and unfit to work or other inactive person*
		7	Fulfilling domestic tasks and care responsibilities
Citizenship	pb220a	1	Austria
		2	EU*
		3	Other*
Personal net income	netIncome		Sum of income components listed below
Employee cash or near cash income	py010n	0	No income
		> 0	Income
Cash benefits or losses from self-employment	py050n	< 0	Losses
		0	No income
		> 0	Benefits
Unemployment benefits	py090n	0	No income
		> 0	Income
Old-age benefits	py100n	0	No income
		> 0	Income
Survivor's benefits	py110n	0	No income
		> 0	Income
Sickness benefits	py120n	0	No income
		> 0	Income
Disability benefits	py130n	0	No income
		> 0	Income
Education-related allowances	py140n	0	No income
		> 0	Income

* combined categories

5. Simulation of close-to-reality population data

Table 5.2: Categorized variables created for use as predictors during the simulation.

Variable	Name	Categories
Age category	ageCat	≤ 15, $(15, 20]$, $(20, 25]$, $(25, 30]$, $(30, 35]$, $(35, 40]$, $(40, 45]$, $(45, 50]$, $(50, 55]$, $(55, 60]$, $(60, 65]$, $(65, 70]$, $(70, 75]$, $(75, 80]$, > 80
Personal net income category	netIncomeCat	$[-9600, -5840]$, $[-5840, -4200)$, $[-4200, 0)$, 0, $(0, 800]$, $(800, 2800]$, $(2800, 5021.56]$, $(5021.56, 8456]$, $(8456, 13720]$, $(13720, 17738]$, $(17738, 23601.65]$ $(23601.65, 29191.86]$, $(29191.86, 36000]$, $(36000, 57227.69]$, > 57227.69

ulated only for persons of age 16 or above (see Eurostat 2004). Furthermore, one category for all persons of age above 80 is used due to the low frequencies of occurrence. In any case, economic status and citizenship are simulated for every region separately. In the multinomial logistic regression models described in Section 5.2.2, the predictors age category, gender and household size are used for economic status, while age category, gender, household size and economic status are then used to simulate citizenship.

In this section, the structure of the simulated categorical variables is evaluated by graphical means only. Figure 5.1 contains mosaic plots visualizing the expected and realized frequencies of gender, region and household size (*top*), as well as gender, economic status and citizenship (*bottom*). Both show very similar structures in the sample and population data. Note that these plots have been selected representatively, as the number of possible combinations of variables is too large to show them all. However, the interactions between all categorical variables are very well reflected in the synthetic population data. This is further documented in Section 5.3.2 by average relative differences of contingency coefficients from multiple simulation runs. While the two plots at the top of Figure 5.1 are nearly identical, closer inspection of the two plots at the bottom reveals small differences. These differences are due to the multinomial logistic regression models. The following two points need to be kept in mind. First, the expected frequencies of the different combinations are solely determined by the sum of the corresponding sample weights. Second, the multinomial models allow simulating combinations that do not occur in the sample but are likely to occur in the population. Consequently, the differences may be interpreted as corrections of the expected frequencies. For additional results concerning the simulation of categorical variables in the case of EU-SILC, including χ^2 goodness of fit tests, the reader is referred to Kraft (2009).

For simulating personal net income, the two approaches described in Section 5.2.3 are compared. In both cases, the variables age category, gender, household size, economic status and citizenship are used as predictors and the models are computed separately for each region.

5.3 Application to EU-SILC

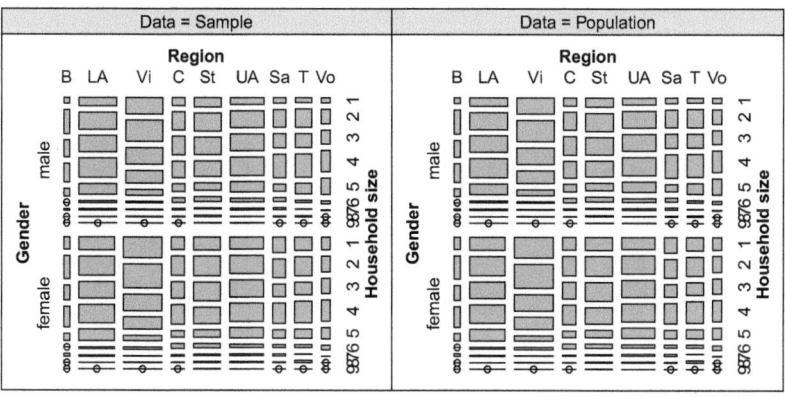

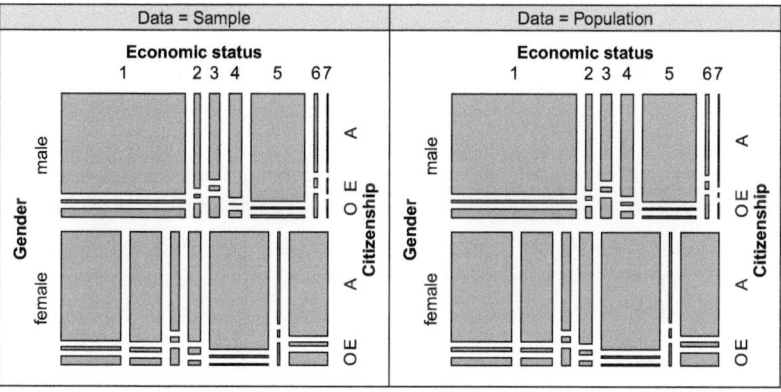

Figure 5.1: *Top*: Mosaic plots of gender, region and household size. *Bottom*: Mosaic plots of gender, economic status and citizenship.

The approach based on multinomial logistic regression models thereby uses the following parameter settings. In the categorization of personal net income, zero is a category of its own since personal net income is a semi-continuous variable. Breakpoints for the positive values are chosen as their weighted 1%, 5%, 10%, 20%, 40%, 60%, 80%, 90%, 95% and 99% quantiles. Furthermore, the only three negative values are used as breakpoints for negative income. See Table 5.2 for the resulting income categories. Values in the categories above the two largest breakpoints are drawn from a truncated generalized Pareto distribution. In the following, this approach will be referred to as *MP*. For the two-step linear regression approach, on the other hand, two different parameter settings are investigated. The first uses random

5. Simulation of close-to-reality population data

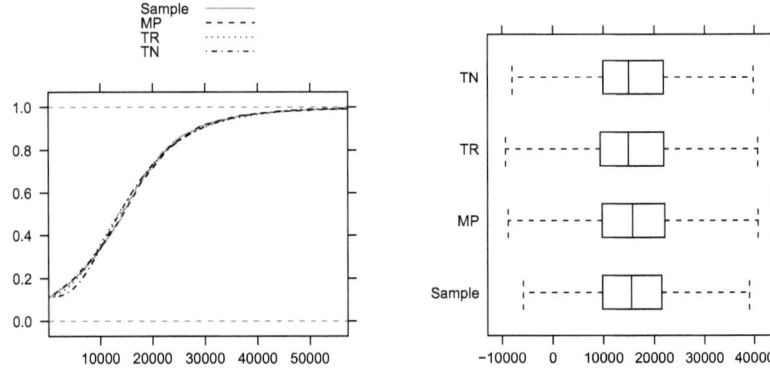

Figure 5.2: *Left*: Cumulative distribution functions of personal net income. For better visibility, the plot shows only the main parts of the data. *Right:* Box plots of personal net income. Points outside the extremes of the whiskers are not plotted.

draws from the residuals and will be referred to as *TR*, the second uses random draws from a normal distribution and will be referred to as *TN*. In both cases, the positive sample data are trimmed with parameters $\alpha_1 = \alpha_2 = 0.01$ and log-transformed in the second step of the procedure. Trimming is used since this performed better (results not shown, cf. Kraft 2009). In order to simulate negative income, a multinomial model is used in the first step. For negative income, again the only three existing values are used as breakpoints (see Table 5.2), and the simulated values are drawn from uniform distributions in the corresponding classes.

In Figure 5.2 (*left*), the cumulative distribution functions (CDF) of personal net income in the three simulated populations are compared to the empirical CDF obtained from the sample. Sample weights are taken into account by adjusting the step height. For better visibility of the differences, the plot shows only the main parts of the data (from 0 to the weighted 99% quantile of the positive values in the sample). The CDFs indicate an excellent fit, in particular for the MP approach. With the TN approach, there are some deviations for lower income, though. Figure 5.2 (*right*) uses box plots to compare the distributions. The box plots are adapted for semi-continuous variables in the following way. Box and whiskers are computed only for the non-zero part of the data and the box widths are proportional to the ratio of non-zero observations to the total number of observed values. For the sample data, sample weights are taken into account when computing the box plot statistics and the box widths. These box plots suggest that the proposed approaches perform well regarding the proportion of individuals with zero income and the distribution of non-zero income for the main part of the data.

5.3 Application to EU-SILC

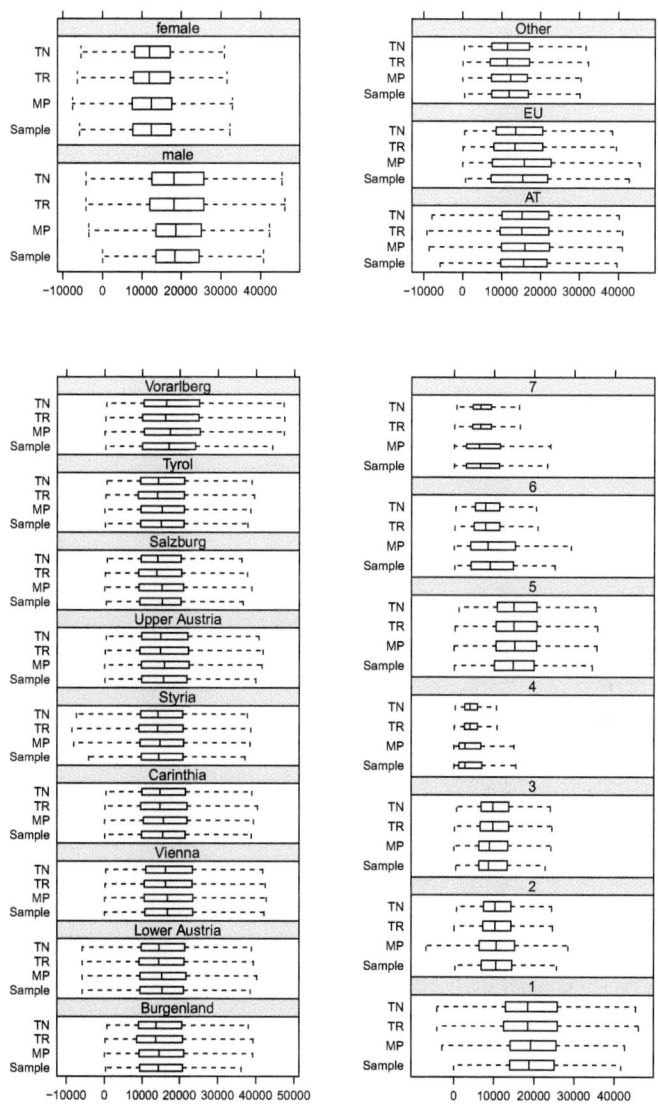

Figure 5.3: Box plots of personal net income split by gender (*top left*), citizenship (*top right*), region (*bottom left*) and economic status (*bottom right*). Points outside the extremes of the whiskers are not plotted.

5. Simulation of close-to-reality population data

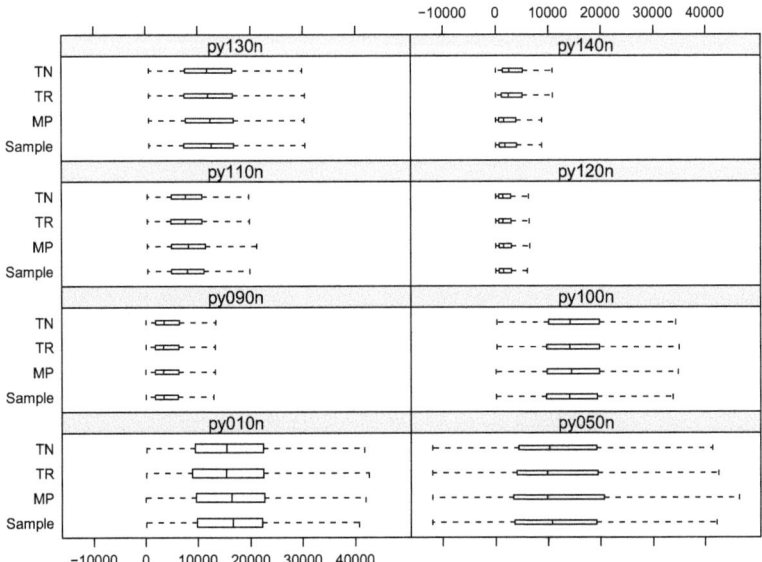

Figure 5.4: Box plots of the income components. Points outside the extremes of the whiskers are not plotted.

Figure 5.3 contains box plots of the conditional distributions of personal net income with respect to gender (*top left*), citizenship (*top right*), region (*bottom left*) and economic status (*bottom right*). The proportions of zeros and the distributions of the non-zero observations appear to be in general well reflected in the simulated populations. Only some very small subgroups of economic status show significant deviations for the two-step regression approaches. This underlines the good fit of the models and illustrates that the proposed methods succeed to account for heterogeneities in the data.

Last but not least, the income components are simulated conditional on income category and economic status (see Section 5.2.4). Box plots of the results are shown in Figure 5.4. Due to the large number of zeros in most income components, a minimum box width is used in some cases to prevent the corresponding boxes from deteriorating into lines. In any case, the plots suggest that the simulation procedure for splitting variables into components works very well.

Additional results from simulations restricted to non-negative income, including correlation coefficients of the income components, can be found in Kraft (2009).

5.3.2 Average results from multiple simulations

In this section, the quality of the proposed methods is further assessed by simulation. With the parameter settings as described in the previous section, 100 populations are simulated. Certain quantities of interest for the sample data are thereby compared to the averages of their population counterparts over all simulation runs. The R package **simFrame** (Alfons et al. 2010c, Alfons 2010) is used to manage the multiple simulations.

The relationships between the categorical variables, including the variables defining the household structure (age, gender and household size), are evaluated using contingency coefficients. Pearson's coefficient of contingency is a measure of association for categorical data defined as $P = \sqrt{\chi^2/(n+\chi^2)}$, where χ^2 is the test statistic of the χ^2 test of independence and n is the number of observations (see, e.g., Kendall and Stuart 1967, for more information).

Furthermore, the proposed methodology for the generation of categorical variables is compared to the framework of Münnich et al. (2003b) and Münnich and Schürle (2003). For the simulation of the household structure, the household sizes in our procedure are obtained in a completely deterministic way by estimated population totals, whereas they draw the household sizes from the observed conditional distributions within the strata. However, the age and gender structure is also generated by resampling households from the corresponding strata. Additional categorical variables are in their framework then simulated by random draws from the observed conditional distributions of the multivariate realizations within each combination of stratum, age category and gender. Keep in mind that this does not allow to simulate combinations that do not occur in the sample.

Table 5.3 compares the contingency coefficients obtained from the sample to the average results over the simulation runs. For the proposed procedure, the relative differences are negligible, except for the coefficient of age and citizenship (pb220a). This exception is a result of using age categories for the prediction of citizenship, which can be avoided by using more categories or the original uncategorized age information. On the other hand, this increases computation time considerably, therefore a reasonable tradeoff has been used. All in all, the correlation structure of the simulated populations is very close to the expected one. For the method of Münnich et al. (2003b) and Münnich and Schürle (2003), the contingency coefficient of age and citizenship also suffers from using age category as conditioning variable. Moreover, the relationships between household size (hsize) and the variables economic status (pl030) and citizenship (pb220a) are not well reflected. This is because these authors suggest to use only stratum, age category and gender as conditioning variables for the simulation of additional categorical variables. Including household size as conditioning variable in the estimation of the conditional multivariate distributions leads to an improvement of the contingency coefficients (results not shown), but causes another problem. Since the size of the sample is rather small, only 3432 of the possible 51030 combinations of region, age category, gender, household size, economic status and citizenship exist in the sample. Hence the

5. Simulation of close-to-reality population data

Table 5.3: Pairwise contingency coefficients of the categorical variables for the sample data (*top*), as well as average results from 100 simulated populations using the proposed method (*middle*), and the method of Münnich et al. (2003b) and Münnich and Schürle (2003) (*bottom*).

		age	rb090	hsize	p1030	pb220a
Sample	db040	0.261	0.019	0.217	0.139	0.161
	age		0.118	0.546	0.723	0.194
	rb090			0.081	0.385	0.026
	hsize				0.404	0.182
	p1030					0.172
Proposed	db040	0.262	0.019	0.217	0.139	0.160
method	age		0.118	0.546	0.716	0.153
	rb090			0.082	0.386	0.026
	hsize				0.405	0.179
	p1030					0.171
Relative	db040	0.283	−2.032	0.000	0.142	−0.073
differences	age		0.021	0.030	−1.098	−21.220
(in %)	rb090			0.339	0.129	−0.580
	hsize				0.108	−1.267
	p1030					−0.445
Method of	db040	0.262	0.019	0.217	0.138	0.160
Münnich et al.	age		0.118	0.546	0.715	0.151
	rb090			0.081	0.386	0.026
	hsize				0.366	0.045
	p1030					0.172
Relative	db040	0.190	−1.139	0.110	−0.536	−0.172
differences	age		0.195	0.037	−1.175	−22.164
(in %)	rb090			−0.288	0.233	2.091
	hsize				−9.423	−74.970
	p1030					−0.055

resulting populations cannot contain any other combinations either. Even though many of the combinations that do not occur are structural zeros, such low variation in the population is simply not realistic as it is very likely that there is a significant number of random zeros resulting from the small sample size (see Section 5.3.3). In short, the proposed method has the advantage that the information from the household size can be included in the simulation of additional variables since the multinomial models allow to simulate combinations that do not exist in the sample.

In Table 5.4, simulated personal net income is evaluated based on various quantities of interest: the percentage of zeros, 5% quantile, median, mean, 95% quantile and standard

Table 5.4: Evaluation of personal net income based on the percentage of zeros, 5% quantile, median, mean, 95% quantile and standard deviation. Values from the sample data are compared to average results from 100 simulated populations.

		%Zeros	5%	Median	Mean	95%	SD
Sample		11.39	2800.00	15428.26	17084.37	36000.00	11589.52
Averages of	MP	11.41	2728.46	15703.79	17135.24	35682.04	11386.47
simulated	TR	11.41	3643.36	14858.96	16980.55	37310.49	11257.74
populations	TN	11.41	4946.36	14949.58	17118.29	36566.82	10296.96
Relative	MP	0.14	−2.55	1.79	0.30	−0.88	−1.75
differences	TR	0.19	30.12	−3.69	−0.61	3.64	−2.86
(in %)	TN	0.19	76.66	−3.10	0.20	1.57	−11.15

deviation. Values from the sample data are compared to the average results for each of the three investigated methods. The relative differences are again used for evaluation. Clearly, the MP approach performs best with an excellent overall fit. For the two-step linear regression procedures, there is considerable deviation in the 5% quantile (cf. Figure 5.2, left). Due to the better fit and the more accurate standard deviation, the TR approach may be favorable over the TN approach.

5.3.3 Influence of sample size and sampling design

In this section, the synthetic population data from Section 5.3.1 are used to evaluate the effect of different sample sizes and sampling designs on the proposed framework in a simulation study. It may not be optimal to use population data that have been generated with the same methodology for such an analysis, but since real population data are not available, this is the only possible way to investigate these issues.

Concerning the sample size, two different scenarios are considered: (i) 6 000 households, which is roughly the real sample size, and (ii) 1% samples, which corresponds to about 35 000 households. In addition, the following two sampling designs are investigated, both of which are frequently used for EU-SILC in practice:

1. Stratified simple random sampling of households by region.

2. Stratified simple random sampling of individuals by region. Then all individuals belonging to the same household as any of the sampled individuals are collected and added to the sample.

The sample sizes were in both cases chosen proportional to the strata sizes. This leads to approximately equal weights for the first design, and weights approximately inverse proportional to the corresponding household sizes for the second design. For each combination of

5. Simulation of close-to-reality population data

Table 5.5: Analysis of empty cells in the contingency tables of the categorical variables. 250 simulated populations are evaluated using the number of empty cells for the initial population (#Initial) and the respective average percentage of false nonempties (%FE), as well as the average number of of additional random empty cells for the samples (#Random) and the respective average proportion of false empties (%FE).

Size	Design	#Initial	%FN	#Random	%FE
Real	1	37730	0.61	10006.48	33.80
Real	2	37730	0.63	9540.84	29.59
1%	1	37730	0.99	6782.76	9.29
1%	2	37730	0.99	6327.52	9.74

sample size and sampling design, 25 samples are drawn from the initial population. Then 10 populations are simulated for each sample, resulting in a total of 250 synthetic populations. Furthermore, calibration using different choices of variables did not have a strong impact on the characteristics of the resulting variables (results not shown). Since households are sampled, however, the resulting sample weights in general do not sum up to the number of individuals in the population. Therefore, calibration on the marginal totals of the regions is performed.

Since the proposed framework allows to simulate combinations of categorical variables that do not occur in the underlying sample, empty cells in the contingency tables are analyzed. Table 5.5 lists the number of empty cells for the initial population (#Initial), the average percentage of these cells that are no longer empty for the simulated populations (false nonempties, %FE), the average number of of additional random empty cells for the samples introduced by the sampling process (#Random), and the average proportion of these cells that are still empty for the simulated populations (false empties, %FE).

For all scenarios, only a very low percentage of combinations that do not exist in the initial population are introduced in the simulated populations. Note that not all empty cells in the contingency table of the initial population are structural zeros. Just because a certain combination does not occur in a specific population does not mean that it is impossible to occur. Thus new combinations introduced in the simulated populations may very well be realistic. In any case, the probability for generating a combination that is in fact a structural zero is very low due to the low percentage of false nonempties.

On the other hand, the large majority of combinations that randomly do not exist in the corresponding sample due to the sampling process are generated in the synthetic populations. Nevertheless, in particular for the small real sample size, there is a considerable amount of such combinations that still do not occur in the simulated populations. The main reason for this is that large households do not occur very frequently in the initial population, hence there is only a low number of such households in the samples, which in turn makes it difficult to reproduce the full variation of possible combinations. This also explains why the first scenario with the real sample size and simple random sampling of households leads to the largest proportion of false empties, as it results in the lowest expected absolute frequencies of large households. To summarize, considering the small sample size for the first two scenarios and the resulting large number of random empty cells in the contingency tables for the samples, the proposed procedure performs quite well.

In Table 5.6, the contingency coefficients between the categorical variables from the initial population are compared to the average results from the simulated populations for each of the four sampling scenarios. For the real sample size, there are considerable differences specifically in the contingency coefficients between the variables region (db040), age and gender (rb090). This is because the household structure is simulated by resampling households from the sample, which due to the small size does not account for all the variation in the initial population. However, since the dependencies within a household are highly complex, the results with the simple resampling approach can still be considered very reasonable. In addition, most of the other relationships are very well reflected. The 1% samples are of course much less affected by the effect of resampling households, and all in all the results are excellent.

Table 5.7 contains an evaluation of the simulated personal net income based on the following quantities of interest: the percentage of zeros, 5% quantile, median, mean, 95% quantile and standard deviation. It should be noted that the reference values for the initial population are computed from the income generated by the MP approach, since this this gave the best fit compared to the original sample data (see Section 5.3.1). The results do not suggest a very strong influence of the sample size or the sampling design and are similar to those from the comparison to the original sample data in Section 5.3.2. For the real sample size, only a small effect of the sampling design on the percentage of zeros is visible in all methods. Furthermore, the sampling design appears to have a slight impact on the two-step linear regression methods in general, most notably on the 5% and 95% quantiles and the standard deviation. In any case, the MP approach clearly gives excellent results and performs best, while the TR method is favorable over the TN method for the two-step approach.

5.4 Conclusions

This paper introduced a flexible framework for simulating population data for household surveys based on available sample data, which is implemented along with diagnostic plots in the R package **simPopulation**. No auxiliary information is used in the procedure, and stratification allows to account for heterogeneities such as regional differences.

Table 5.6: Pairwise contingency coefficients of the categorical variables for the initial population, as well as average results from 250 simulated populations for each of the four sampling scenarios.

		age	rb090	hsize	p1030	pb220a
Population	db040	0.261	0.020	0.217	0.138	0.160
	age		0.118	0.546	0.716	0.153
	rb090			0.082	0.386	0.026
	hsize				0.405	0.179
	p1030					0.172
Real size,	db040	0.337	0.025	0.256	0.153	0.162
Design 1	age		0.141	0.565	0.717	0.162
	rb090			0.086	0.387	0.029
	hsize				0.408	0.186
	p1030					0.174
Relative	db040	29.055	28.335	18.069	10.265	0.740
differences	age		19.209	3.512	0.209	6.091
(in %)	rb090			5.692	0.355	8.711
	hsize				0.835	3.952
	p1030					1.316
Real size	db040	0.347	0.028	0.239	0.157	0.165
Design 2	age		0.142	0.560	0.716	0.162
	rb090			0.080	0.388	0.027
	hsize				0.404	0.188
	p1030					0.176
Relative	db040	32.810	41.592	10.129	13.667	3.091
differences	age		20.100	2.631	0.121	6.142
(in %)	rb090			−1.475	0.592	2.197
	hsize				−0.097	4.590
	p1030					2.664
1% sample,	db040	0.278	0.020	0.223	0.141	0.161
Design 1	age		0.123	0.549	0.716	0.153
	rb090			0.082	0.385	0.027
	hsize				0.406	0.182
	p1030					0.171
Relative	db040	6.352	1.480	2.951	1.640	0.577
differences	age		4.048	0.611	0.029	0.524
(in %)	rb090			0.532	−0.182	1.185
	hsize				0.308	1.212
	p1030					−0.062
1% sample,	db040	0.277	0.021	0.221	0.140	0.161
Design 2	age		0.122	0.549	0.716	0.154
	rb090			0.082	0.386	0.026
	hsize				0.406	0.179
	p1030					0.171
Relative	db040	6.024	4.893	1.775	1.067	0.207
differences	age		3.720	0.593	0.020	0.755
(in %)	rb090			0.101	0.138	0.559
	hsize				0.196	−0.065
	p1030					−0.525

5.4 Conclusions

Table 5.7: Evaluation of personal net income based on the percentage of zeros, 5% quantile, median, mean, 95% quantile and standard deviation. Values from the initial population are compared to average results from 250 simulated populations for each of the four sampling scenarios.

		%Zeros	5%	Median	Mean	95%	SD
Population		11.40	2719.73	15700.65	17130.72	35677.48	11390.32
Real size,	MP	11.26	2669.72	15667.54	17064.57	35601.94	11232.09
Design 1	TR	11.28	3514.98	14770.39	16900.82	37305.08	11209.41
	TN	11.28	4755.06	14990.06	17377.93	38028.43	10935.30
Relative	MP	−1.15	−1.84	−0.21	−0.39	−0.21	−1.39
differences	TR	−1.00	29.24	−5.92	−1.34	4.56	−1.59
(in %)	TN	−1.00	74.84	−4.53	1.44	6.59	−3.99
Real size,	MP	11.44	2708.81	15665.23	17136.36	35826.75	11385.18
Design 2	TR	11.45	3397.59	14879.76	17097.54	38101.73	11540.92
	TN	11.45	4743.06	15095.90	17610.83	38929.09	11278.78
Relative	MP	0.38	−0.40	−0.23	0.03	0.42	−0.05
differences	TR	0.45	24.92	−5.23	−0.19	6.79	1.32
(in %)	TN	0.45	74.39	−3.85	2.80	9.11	−0.98
1% sample,	MP	11.37	2720.79	15695.71	17113.44	35643.94	11279.51
Design 1	TR	11.37	3529.14	14834.87	16948.50	37324.55	11196.36
	TN	11.37	4838.43	15049.16	17434.75	38057.47	10907.39
Relative	MP	−0.22	0.04	−0.03	−0.10	−0.09	−0.97
differences	TR	−0.19	29.76	−5.51	−1.06	4.62	−1.70
(in %)	TN	−0.19	77.90	−4.15	1.77	6.67	−4.24
1% sample,	MP	11.36	2723.38	15699.97	17134.05	35661.53	11340.96
Design 2	TR	11.37	3406.91	14913.92	17085.76	37937.72	11455.55
	TN	11.37	4810.99	15134.27	17628.18	38840.56	11211.61
Relative	MP	−0.31	0.13	−0.00	0.02	−0.04	−0.43
differences	TR	−0.25	25.27	−5.01	−0.26	6.34	0.57
(in %)	TN	−0.25	76.89	−3.61	2.90	8.87	−1.57

The proposed framework is applicable to a broad class of surveys and led to excellent results in an application to EU-SILC. For simulation of personal net income, using multinomial models combined with random draws from the resulting categories and generalized Pareto tail modeling performed better than two-step regression, but is computationally more expensive. The computation time of the multinomial models thereby strongly depends on the number of categories used in the discretization. Concerning the two-step approach, trimming combined with random draws from the residuals appeared to be favorable. Nevertheless, the choice of method also depends on the purpose. For simulation studies in survey statistics, it is important not to favor any of the investigated methods by the underlying data generation procedure in order to avoid biased simulation results.

Acknowledgements The authors are grateful to the referees for helpful comments and suggestions.

Chapter 6

Disclosure risk of synthetic population data with application in the case of EU-SILC[1]

Published in *Privacy in Statistical Databases*, volume 6344 of *Lecture Notes in Computer Science* (Templ and Alfons 2010).

Matthias Templ[a,b], Andreas Alfons[a]

[a] Department of Statistics and Probability Theory, Vienna University of Technology
[b] Methods Unit, Statistics Austria

Abstract In survey statistics, simulation studies are usually performed by repeatedly drawing samples from population data. Furthermore, population data may be used in courses on survey statistics to support the theory by practical examples. However, real population data containing the information of interest are in general not available, therefore synthetic data need to be generated. Ensuring data confidentiality is thereby absolutely essential, while the simulated data should be as realistic as possible. This paper briefly outlines a recently proposed method for generating close-to-reality population data for complex (household) surveys, which is applied to generate a population for Austrian EU-SILC (European Union Statistics on Income and Living Conditions) data. Based on this synthetic population, confidentiality issues are discussed using five different worst case scenarios. In all scenarios, the intruder has the complete information on key variables from the real survey data. It is shown that even in these worst case scenarios the synthetic population data are confidential. In addition, the synthetic data are of high quality.

---
[1] This work was partly funded by the European Union (represented by the European Commission) within the 7th framework programme for research (Theme 8, Socio-Economic Sciences and Humanities, Project AMELI (Advanced Methodology for European Laeken Indicators), Grant Agreement No. 217322). For more information on the project, visit http://ameli.surveystatistics.net.

# 6. Disclosure risk of synthetic population data

Keywords Survey Statistics, Synthetic Population Data, Data Confidentiality

6.1 Introduction

In the analysis of survey data, variability due to sampling, imputation of missing values, measurement errors and editing must be considered. Statistical methods thus need to be evaluated with respect to the effect of these variabilities on point and variance estimates. A frequently used strategy to adequately measure such effects under different settings is to perform simulation studies by repeatedly drawing samples from population data (possibly using different sampling designs) and to compare the estimates with the true values of the sample frame. Evaluating and comparing various statistical methods within such a *design-based* simulation approach under different *close-to-reality* settings is daily work for survey statisticians and has been done, e.g., in the research projects DACSEIS (Münnich et al. 2003b), EurEdit (Chambers 2001) and AMELI (Alfons et al. 2009).

Furthermore, population data may be used for teaching courses on survey statistics. Realistic examples may help students to better understand issues in survey sampling, e.g., regarding different sampling designs.

Since suitable population data are typically not available, it is necessary to generate synthetic data. The generation of population microdata for selected surveys as a basis for Monte Carlo simulation studies is described in Münnich et al. (2003b), Münnich and Schürle (2003). These procedures were extended in Alfons et al. (2009, 2010b) to simulate close-to-reality population data for more complex surveys such as EU-SILC (*European Union Statistics on Income and Living Conditions*). However, confidentiality issues of such synthetic population data are only briefly addressed in these contributions.

Generation of population microdata for simulation studies is closely related to the field of *microsimulation* (Clarke 1996), yet the aims are quite different. Microsimulation models attempt to reproduce the behavior of individual units within a population for policy analysis purposes and are well-established within the social sciences. Nevertheless, they are highly complex and time-consuming. On the other hand, synthetic population microdata for simulation studies in survey statistics are used to evaluate the behavior of statistical methods. Thus fast computation is preferred to over-complex models.

Another approach towards the generation of microdata is to use multiple imputation to create *fully* or *partially* synthetic data sets, as proposed in Rubin (1993), Little (1993). This approach is further discussed in Raghunathan et al. (2003), Drechsler et al. (2008), Reiter (2009). However, these contributions do not consider some important issues such as the generation of categories that do not occur in the original sample or the generation of structural zeros.

The rest of the paper is organized as follows. Section 6.2 outlines the framework for generating synthetic populations proposed in Alfons et al. (2010b). Then the data investigated in this paper are introduced in Section 6.3. Sections 6.4 and 6.5 discuss statistical disclosure control issues related to survey and population data. In Section 6.6, several scenarios for evaluating the confidentiality of synthetic population data are described, while Section 6.7 lists the obtained results for these scenarios. The final Section 6.8 concludes.

6.2 Generation of synthetic population data

The generation of synthetic population data for surveys is described in great detail in Alfons et al. (2010b). Therefore, only the basic ideas of this framework are presented here. Several conditions for simulating population data are listed in Münnich et al. (2003b), Münnich and Schürle (2003), Alfons et al. (2010b). The most important requirements are:

- Actual sizes of regions and strata need to be reflected.

- Marginal distributions and interactions between variables should be represented accurately.

- Heterogeneities between subgroups, in particular regional aspects, should be allowed.

- Data confidentiality must be ensured.

In general, the framework for generating synthetic population data consists of four steps:

1. In case of household data, set up the household structure.

2. Simulate categorical variables.

3. Simulate continuous variables.

4. Split continuous variables into components.

Not all of these steps need to be performed, depending on the survey of interest.

Step 1. When generating household data, the household structure is simulated separately for the different household sizes within each strata. Using the corresponding sample weights, the number of households is simply estimated by the Horvitz-Thompson estimator (Horvitz and Thompson 1952). The structure of the population households is then simulated by resampling some basic variables from the sample households with probabilities proportional to the sample weights. For disclosure reasons, information from as few variables as possible should be used to construct the household structure (e.g., only age and gender information).

6. Disclosure risk of synthetic population data

Step 2. For each stratum, the conditional distribution of any additional categorical variable is estimated with a multinomial logistic regression model. The previously simulated variables are thereby used as predictors. Furthermore, the sample weights are considered and it is possible to account for structural zeros. The main advantage of this approach is that it allows to generate combinations that do not occur in the sample, which is not the case for the procedure introduced in Münnich et al. (2003b), Münnich and Schürle (2003).

Step 3. Two approaches for simulating continuous variables are proposed in Alfons et al. (2010b), but only the approach that performs better in the case of EU-SILC data is considered in this paper. First, the variable to be simulated is discretized using suitable breakpoints. The discretized variable is then then simulated as described in the previous step. Finally, the values of the continuous variable are randomly drawn from uniform distributions within the respective intervals. Note that the idea behind this approach is to divide the data into relatively small subsets so that the uniform distribution is not too much of an oversimplification.

Step 4. Splitting continuous variables into components is based on conditional resampling of fractions from the sample households with probabilities proportional to the sample weights. Only very few highly influential categorical variables should thereby be considered for conditioning. The resampled fractions are then multiplied with the previously simulated total.

The data simulation framework proposed in Alfons et al. (2010b) is implemented in the R (R Development Core Team 2010) package **simPopulation** (Alfons and Kraft 2010). In addition to the four steps of the procedure and a wrapper function to generate synthetic EU-SILC populations, various diagnostic plots are available.

6.3 Synthetic EU-SILC population data

The *European Union Statistics on Income and Living Conditions* (EU-SILC) is a complex panel survey conducted in EU member states and other European countries. It is mainly used for measuring risk-of-poverty and social cohesion in Europe (Atkinson et al. 2002). The generation of synthetic population data based on Austrian EU-SILC survey data from 2006 is discussed and evaluated in Alfons et al. (2010b). The resulting synthetic population is investigated in this paper with respect to confidentiality issues. Table 6.1 lists the variables that are used in the analysis. A detailed description of all variables included in EU-SILC data and their possible outcomes is given in Eurostat (2004).

6.3 Synthetic EU-SILC population data

Table 6.1: Variables of the synthetic EU-SILC population data used in this paper.

Variable	Type	
Region	Categorical	9 levels
Household size	Categorical	9 levels
Age category	Categorical	15 levels
Gender	Categorical	2 levels
Economic status	Categorical	7 levels
Citizenship	Categorical	3 levels
Personal net income	Semi-continuous	

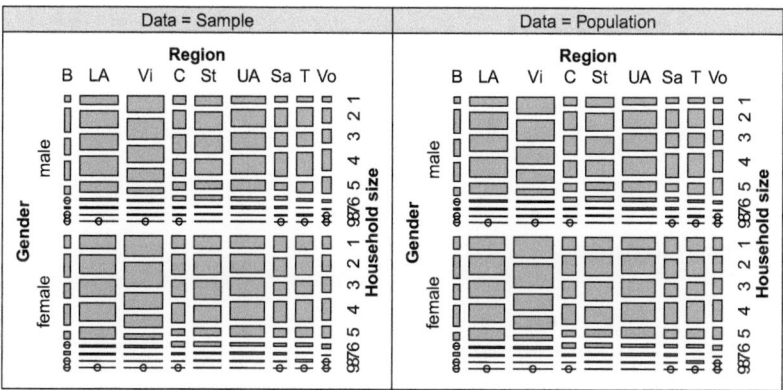

Figure 6.1: Mosaic plots of gender, region and household size of the Austrian EU-SILC sample from 2006 and the resulting synthetic population.

In order to demonstrate that the synthetic population data are of high quality, they are compared to the underlying sample data. Figure 6.1 contains mosaic plots of gender, region and household size for the sample and synthetic population data, respectively. Clearly, the plots show almost identical structures. In addition, the distribution of personal net income is visualized in Figure 6.2. On the left hand side, the cumulative distribution functions for the sample and population data, respectively, are displayed. For better visibility, only the main parts of the data are shown, which are nearly in perfect superposition. On the right hand side, the conditional distributions with respect to gender are represented by box plots. The fit of the distribution within the subgroups is excellent and heterogeneities between the subgroups are very well reflected. Note that points outside the extremes of the whiskers are not plotted. For extensive collections of results showing that the multivariate structure of the data is well preserved, the reader is referred to Alfons et al. (2010b), Kraft (2009).

6. Disclosure risk of synthetic population data

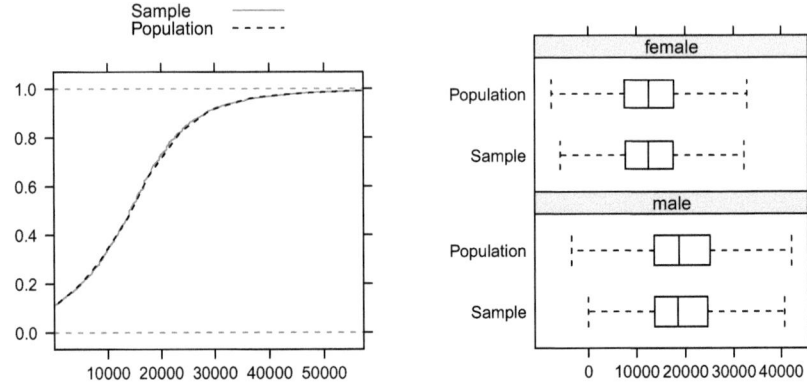

Figure 6.2: Personal net income in the Austrian EU-SILC sample from 2006 and the resulting synthetic population. *Left:* Cumulative distribution functions of personal net income. Only the main parts of the data are shown for better visibility. *Right:* Box plots of the conditional distributions with respect to gender. Points outside the extremes of the whiskers are not plotted.

6.4 A global disclosure risk measure for survey data

A popular global measure of the reidentification risk for survey data is given by the number of uniquenesses in the sample that are unique in the population as well. Let m categorical key variables in the sample and population data be denoted by $\boldsymbol{x}_j^S = (x_{1j}^S, \ldots, x_{nj}^S)'$ and $\boldsymbol{x}_j^P = (x_{1j}^P, \ldots, x_{Nj}^P)'$, respectively, $j = 1, \ldots, m$, where n and N give the corresponding number of observations. For an observation in the sample given by the index $c = 1, \ldots, n$, let J_c^S and J_c^P denote the index sets of observations in the sample and population data, respectively, with equal values in the m key variables:

$$\begin{aligned} J_c^S &:= \{j = 1, \ldots, n : x_{jk}^S = x_{ck}^S, \; k = 1, \ldots, m\}, \\ J_c^P &:= \{j = 1, \ldots, N : x_{jk}^P = x_{ck}^S, \; k = 1, \ldots, m\}. \end{aligned} \quad (6.1)$$

Furthermore, a function $\mathcal{I}$ is defined as

$$\mathcal{I}(J) := \begin{cases} 1 & \text{if } |J| = 1, \\ 0 & \text{else.} \end{cases} \quad (6.2)$$

The global disclosure risk measure can then be expressed by

$$\tau_0 := \sum_{c=1}^{n} \mathcal{I}(J_c^S) \cdot \mathcal{I}(J_c^P). \tag{6.3}$$

Note that the notation in (6.3) differs from the common definition. For comparison, see, e.g., the risk measures in Rinott and Shlomo (2006), Elamir and Skinner (2006). The notation used in (6.3) describes the same phenomenon, but provides more flexibility in terms of software implementation (Templ et al. 2009) and allows to formulate the adapted risk measures given in the following section.

Clearly, the risk of reidentification is lower the higher the corresponding population frequency count. If the population frequency count is sufficiently high, it is not possible for an intruder to assign the observation for which they hold information with absolute certainty. Hence the intruder does not know whether the reidentification was successful. However, the true frequency counts of the population are usually unknown and need to be estimated by modeling the distribution of the population frequencies.

In Section 6.6, the global disclosure risk measure τ_0 is modified to estimate the disclosure risk for synthetic population data in certain scenarios instead of survey data.

6.5 Confidentiality of synthetic population data

The motivation for generating close-to-reality population data is to make the resulting data sets publicly available to researchers for use in simulation studies or courses on survey statistics. Therefore, the disclosure risk of such data needs to be low, while at the same time the multivariate structure should be as realistic as possible.

If population data are generated from perturbed survey data, confidentiality is guaranteed whenever the underlying survey data are confidential. Perturbing survey data is typically done by performing recodings and suppression such that k-anonymity (Samerati and Sweeney 1998, Sweeney 2002) is provided for categorical key variables, as well as low risk of reidentification on the individual level is ensured (Franconi and Polettini 2004, Domingo-Ferrer and Mateo-Sanz 2002, Templ and Meindl 2008, and references therein). In any case, perturbation implies information loss. Usually not all combinations of categorical key variables are still included in the perturbed sample and outliers in continuous variables are often modified to a great extent. It is thus favorable to use information of the unperturbed sample to generate synthetic populations, as this increases the quality of the resulting data.

Based on the ideas proposed in Rubin (1993), Little (1993), the generation of *fully* or *partially* synthetic population data using multiple imputation is discussed in Raghunathan et al. (2003), Drechsler et al. (2008), Reiter (2009). More precisely, let p be the number of variables in the sample and let the first k, $1 \leq k < p$, categorical variables be available for the

6. Disclosure risk of synthetic population data

population data from administrative sources. These first k variables are released unchanged, while the remaining $p-k$ variables are estimated using regression based multiple imputation. It is important to note that the first k variables of real population data may still contain unique combinations after cross tabulation, therefore further investigation may be necessary to ensure confidentiality. Probabilities of reidentification for such synthetic data have been studied in Reiter and Mitra (2009), based on the work of Duncan and Lambert (1986), Fienberg et al. (1997), by matching the synthetic data with the intruder's data on some predefined key variables.

The situation for synthetic population data generated by the approach of Alfons et al. (2010b) is somewhat different. A very low number of basic categorical variables are generated in the first step by resampling from the actual survey data. Since the sample weights are thereby used as probability weights, on average k-anonymity is provided with respect to these basic variables, where k denotes the smallest sample weight. In surveys, k is typically very high (> 500), hence the disclosure risk is very low. However, additional categorical and continuous variables are generated based on models obtained from the actual survey data. In particular, the generation of continuous variables involves random draws from certain intervals.

With the additional categorical variables, some unique combinations may be introduced in the synthetic population data. If such a combination is not unique in the real population, it is not useful to an intruder. On the other hand, if such a combination is unique in the real population as well, it must be ensured that the values of the other variables in the synthetic population data are not too close to the real values. Most notably, it is of interest to measure the difference in continuous variables of the successfully matched statistical units.

In addition, unique combinations in the real population may even be critical if they are not unique in the synthetic population data. An intruder could in this case look for all occurrences of such a combination in the synthetic population. If the corresponding units have too similar values in a (continuous) variable of interest, the intruder may be able to infer some information on the original value, since the synthetic values have been predicted with models obtained from the real sample data.

In order to investigate these issues in more detail, various disclosure scenarios are introduced in the following section. Section 6.7 then presents the results for the synthetic EU-SILC population data described in Section 6.3.

6.6 Disclosure scenarios for synthetic population data

Five different scenarios are considered to evaluate the confidentiality of synthetic data generated with the framework proposed in Alfons et al. (2010b). These scenarios are motivated by the synthetic EU-SILC population data, hence only a continuous variable is considered

6.6 Disclosure scenarios for synthetic population data

to contain confidential information, while there are m categorical key variables. In the case of EU-SILC, the confidential variable is *personal net income* and the key variables are *region, household size, age category, gender, economic status* and *citizenship* (see Table 6.1). Let the confidential continuous variable for the original sample and synthetic population, respectively, be denoted by $\boldsymbol{y}^S = (y_1^S, \ldots, y_n^S)'$ and $\boldsymbol{y}^U = (y_1^U, \ldots, y_N^U)'$, while the categorical key variables are denoted by $\boldsymbol{x}_j^S = (x_{1j}^S, \ldots, x_{nj}^S)'$ and $\boldsymbol{x}_j^U = (x_{1j}^U, \ldots, x_{Nj}^U)'$, $j = 1, \ldots, m$, analogous to the definitions in Section 6.4. Furthermore, let J_c^S be defined as in (6.1) and let J_c^U be defined accordingly as

$$J_c^U := \{j = 1, \ldots, N : x_{jk}^U = x_{ck}^S, \ k = 1, \ldots, m\}. \tag{6.4}$$

In the following scenarios, the intruder has knowledge of the m key variables for all observations from the original sample and tries to acquire information on the confidential variable.

It should be noted that the link to the global risk measure from (6.3) is loosened in the following. Disclosure is considered to occur if the value of the confidential variable for a unique combination of key variables in the sample can be closely estimated from the synthetic population data, given a prespecified value of accuracy p. However, such a sample uniqueness does not need to be unique in the true population, in which case close estimation of the confidential variable would not necessarily result in disclosure. In this sense, the following scenarios can be considered worst case scenarios and the reidentification risk is thus overestimated. Proper analysis with estimation of the true population uniquenesses is future work.

6.6.1 Scenario 1: Attack using one-to-one matches in key variables with information on the data generation process

The intruder in this scenario tries to find one-to-one matches between their data and the synthetic population data. Moreover, they know the intervals from which the synthetic values were drawn. For details on the data generation procedure, the reader is referred to Alfons et al. (2010b). Let these intervals be denoted by $[l_j, u_j]$, $j = 1, \ldots, N$, and let l be a function giving the length of an interval defined as $l([a,b]) := b - a$ and $l(\emptyset) := 0$. With a prespecified value of accuracy p defining a symmetric interval around a confidential value, (6.3) is reformulated as

$$\tau_1 := \sum_{c=1}^{n} \mathcal{I}(J_c^S) \cdot \mathcal{I}(J_c^U) \cdot \frac{l([y_c^S(1-p), y_c^S(1+p)] \cap [l_{j_c}, u_{j_c}])}{l([l_{j_c}, u_{j_c}])}, \tag{6.5}$$

where $j_c \in J_c^U$ if $|J_c^U| = 1$, i.e., j_c is the index of the unit in the synthetic population with the same values in the key variables as the cth unit in the intruder's data if such a one-to-one match exists, otherwise it is a dummy index. The last term in (6.5) thereby gives the probability that for the successfully matched unit, the synthetic value drawn from the interval $[l_{j_c}, u_{j_c}]$ is sufficiently close to the original value y_c^S.

6.6.2 Scenario 2: Attack using one-to-one matches in key variables without information on the data generation process

In general, an intruder does not have any knowledge on the intervals from which the synthetic values were drawn. In this case, reidentification is successful if the synthetic value itself of a successfully matched unit is sufficiently close to the original value. The risk of reidentification thus needs to be reformulated as

$$\tau_2 := \sum_{c=1}^{n} \mathcal{I}(J_c^S) \cdot \mathcal{I}(J_c^U) \cdot \mathbb{I}_{[y_c^S(1-p), y_c^S(1+p)]}(y_{j_c}^U), \qquad (6.6)$$

where j_c is defined as above and $\mathbb{I}_A$ denotes the indicator function for a set A.

6.6.3 Scenario 3: Attack using all occurrences in key variables with information on the data generation process

This scenario is an extension of Scenario 1, in which the intruder does not only try to find one-to-one matches, but looks for all occurrences of a unique combination from their data in the synthetic population data. Keep in mind that the intruder in this scenario knows the intervals from which the synthetic values were drawn. For a unique combination in the intruder's data, reidentification is possible if the probability that the synthetic values of all matched units are sufficiently close to the original value. Hence the disclosure risk from (6.5) changes to

$$\tau_3 := \sum_{c=1}^{n} \mathcal{I}(J_c^S) \cdot \prod_{j \in J_c^U} \frac{l([y_c^S(1-p), y_c^S(1+p)] \cap [l_j, u_j])}{l([l_j, u_j])}. \qquad (6.7)$$

6.6.4 Scenario 4: Attack using all occurrences in key variables without information on the data generation process

In an analogous extension of Scenario 2, reidentification of a unique combination from the intruder's data is successful if the synthetic values themselves of all matched units are sufficiently close to the original value. Equation (6.6) is in this case rewritten as

$$\tau_4 := \sum_{c=1}^{n} \mathcal{I}(J_c^S) \cdot \prod_{j \in J_c^U} \mathbb{I}_{[y_c^S(1-p), y_c^S(1+p)]}(y_j^U). \qquad (6.8)$$

6.6.5 Scenario 5: Attack using key variables for model predictions

In this scenario, the intruder uses the information from the synthetic population data to obtain a linear model for $\boldsymbol{y}^U$ with predictors $\boldsymbol{x}_j^U$, $j = 1, \ldots, m$:

$$\boldsymbol{y}^U = \beta_0 + \beta_1 \boldsymbol{x}_1^U + \ldots + \beta_m \boldsymbol{x}_m^U + \boldsymbol{\varepsilon}. \qquad (6.9)$$

For a unique combination of the key variables, reidentification is possible if the corresponding predicted value is sufficiently close to the original value. Let the predicted values of the intruder's data be denoted by $\hat{\boldsymbol{y}}^S = (\hat{y}_1^S, \ldots, \hat{y}_n^S)'$. Then the disclosure risk can be formulated as

$$\tau_5 := \sum_{c=1}^{n} \mathcal{I}(J_c^S) \cdot \mathbb{I}_{[y_c^S(1-p), y_c^S(1+p)]}(\hat{y}_c^S). \qquad (6.10)$$

Note that for large population data, the computational costs for fitting such a regression model are very high, so an intruder needs to have a powerful computer with very large memory. Furthermore, the intruder could also perform a stepwise model search using an optimality criterion such as the AIC (Akaike 1970).

6.7 Results

The disclosure risk of the synthetic Austrian EU-SILC population data described in Section 6.3 is analyzed in the following with respect to the scenarios defined in the previous section. In these scenarios, the intruder has knowledge of the categorical key variables *region*, *household size*, *age category*, *gender*, *economic status* and *citizenship* for all observations in the original sample used to generate the data. In addition, the intruder tries to obtain information on the confidential variable *personal net income* (see Table 6.1 for a description of these variables). The original sample thereby consists of $n = 14\,883$ and the synthetic population of $N = 8\,182\,218$ observations.

Note that this paper only evaluates the risk of reidentification for this specific synthetic data set. In order to get more general results regarding confidentiality of the data generation process, many data sets need to be generated in a simulation study and the average values need to be reported. This task, however, is beyond the scope of this paper.

Table 6.2 lists the results for the risk measures for the investigated scenarios using different values of the accuracy parameter p. Besides the absolute values, the relative values with

6. Disclosure risk of synthetic population data

Table 6.2: Results for Scenarios 1-5 using different values for the accuracy parameter p.

Scenario	Risk measure	p = 0.01	0.02	0.05
1	τ_1	0	0	0.052
2	τ_2	0	0	0
3	τ_3	$1.1 \cdot 10^{-8}$	$1.2 \cdot 10^{-6}$	0.053
4	τ_4	15	15	15
5	τ_5	20	43	110
1	τ_1/n	0	0	$3.5 \cdot 10^{-6}$
2	τ_2/n	0	0	0
3	τ_3/n	$6.7 \cdot 10^{-13}$	$8.6 \cdot 10^{-11}$	$3.5 \cdot 10^{-6}$
4	τ_4/n	0.001	0.001	0.001
5	τ_5/n	0.001	0.003	0.007

respect to the size of the intruder's data set are presented, which give the probabilities of successful reidentification.

The results show that even if an intruder is able to reidentify an observation, they do not gain any useful information, as the probability that the obtained value is sufficiently close to the original value is extremely low.

In particular if the intruder tries to find one-to-one matches (Scenarios 1 and 2), the probability of a successful reidentification is only positive for $p = 0.05$ and if they have information on the data generation process, i.e., the intervals from which the synthetic values were drawn.

If the intruder looks for all occurrences of a unique combination from their data in the synthetic population, using information on the data generation process hardly changes the probabilities of reidentification (Scenario 3). This is not a surprise given the formula in (6.7), since for such a unique combination, the probabilities that the corresponding synthetic values are sufficiently close to the original value need to be multiplied. On the other hand, if the intruder uses only the synthetic values (Scenario 4), some observations are successfully reidentified. Nevertheless, the probabilities of reidentification are extremely low.

Among the considered scenarios, Scenario 5 leads to the highest disclosure risk. However, the regression model in this scenario comes with high computational costs and the probabilities of reidentification are still far too low to obtain any useful information.

6.8 Conclusions

Synthetic population data play an important part in the evaluation of statistical methods in the survey context. Without such data, it is not possible to perform design-based simulation studies.

6.8 Conclusions

This paper gives a brief outline of the flexible framework proposed in Alfons et al. (2010b) for simulating population data for (household) surveys based on available sample data. The framework is applicable to a broad class of surveys and is implemented along with diagnostic plots in the R package **simPopulation**. In the case of EU-SILC, the data generation procedure led to excellent results with respect to information loss.

In this paper, confidentiality issues of the generated synthetic EU-SILC population are discussed based on five different worst case scenarios. The results show that while reidentification is possible, an intruder would not gain any useful information from the purely synthetic data. Even if they successfully reidentify a unique combination of key variables from their data, the probability that the obtained value is close to the original value is extremely low for all considered worst case scenarios.

Due to our experiences and the results from the investigated scenarios, we can argue that synthetic population data generated with the methodology introduced in Alfons et al. (2010b) and implemented in **simPopulation** are confidential and can be distributed to the public. Researchers could then use this data to evaluate the effects of different sampling designs, missing data mechanisms and outlier models on the estimator of interest in design-based simulation studies.

6. Disclosure risk of synthetic population data

Chapter 7

A comparison of robust methods for Pareto tail modeling in the case of Laeken indicators

Slightly corrected reprint of Alfons et al. (2010e), which has been published in *Combining Soft Computing and Statistical Methods in Data Analysis*, volume 77 of *Advances in Intelligent and Soft Computing*.

Andreas Alfons[a], Matthias Templ[a,b], Peter Filzmoser[a], Josef Holzer[a,c]

[a] Department of Statistics and Probability Theory, Vienna University of Technology
[b] Methods Unit, Statistics Austria
[c] now at Landesstatistik Steiermark

Abstract The Laeken indicators are a set of indicators for measuring poverty and social cohesion in Europe. However, some of these indicators are highly influenced by outliers in the upper tail of the income distribution. This paper investigates the use of robust Pareto tail modeling to reduce the influence of outlying observations. In a simulation study, different methods are evaluated with respect to their effect on the quintile share ratio and the Gini coefficient.

7.1 Introduction

As a monitoring system for policy analysis purposes, the European Union introduced a set of indicators, called the *Laeken indicators*, to measure risk-of-poverty and social cohesion in Europe. The basis for most of these indicators is the EU-SILC *(European Union Statistics on Income and Living Conditions)* survey, which is an annual panel survey conducted in EU member states and other European countries. Most notably for this paper, EU-SILC data contain information on the income of the sampled households. Each person of a household is thereby assigned the same *equivalized disposable income* (EU-SILC 2004). The subset of Laeken indicators based on EU-SILC is computed from this equivalized income, taking into account the sample weights.

In general the upper tail of an income distribution behaves differently than the rest of the data and may be modeled with a *Pareto* distribution. Moreover, EU-SILC data typically contain some extreme outliers that not only have a strong influence on some of the Laeken indicators, but also on fitting the Pareto distribution to the tail. Modeling the tail in a robust manner should therefore improve the estimates of the affected indicators.

The rest of the paper is organized as follows. Section 7.2 gives a brief description of selected Laeken indicators, while Section 7.3 discusses Pareto tail modeling. A simulation study is performed in Section 7.4 and Section 7.5 concludes.

7.2 Selected Laeken indicators

This paper investigates the influence of promising robust methods for Pareto tail modeling on the *quintile share ratio* and the *Gini coefficient*. Both indicators are measures of inequality and are highly influenced by outliers in the upper tail. Strictly following the Eurostat definitions (EU-SILC 2004), the indicators are implemented in the R package **laeken** (Alfons et al. 2010a).

For the following definitions, let $x := (x_1, \ldots, x_n)'$ be the equivalized disposable income with $x_1 \leq \ldots \leq x_n$ and let $w := (w_i, \ldots, w_n)'$ be the corresponding personal sample weights, where n denotes the number of observations.

7.2.1 Quintile share ratio

The income quintile share ratio is defined as the ratio of the sum of equivalized disposable income received by the 20% of the population with the highest equivalized disposable income to that received by the 20% of the population with the lowest equivalized disposable income (EU-SILC 2004). Let $q_{0.2}$ and $q_{0.8}$ denote the weighted 20% and 80% quantiles of x with weights w, respectively. With $I_{\leq q_{0.2}} := \{i \in \{1, \ldots, n\} : x_i \leq q_{0.2}\}$ and $I_{>q_{0.8}} := \{i \in \{1, \ldots, n\} : x_i > q_{0.8}\}$, the quintile share ratio is estimated by

$$QSR := \frac{\sum_{i \in I_{>q_{0.8}}} w_i x_i}{\sum_{i \in I_{\leq q_{0.2}}} w_i x_i}. \tag{7.1}$$

7.2.2 Gini coefficient

The Gini coefficient is defined as the relationship of cumulative shares of the population arranged according to the level of equivalized disposable income, to the cumulative share of the equivalized total disposable income received by them (EU-SILC 2004). In mathematical terms, the Gini coefficient is estimated by

$$Gini := 100 \left[\frac{2 \sum_{i=1}^{n} \left(w_i x_i \sum_{j=1}^{i} w_j \right) - \sum_{i=1}^{n} w_i^2 x_i}{\left(\sum_{i=1}^{n} w_i \right) \sum_{i=1}^{n} (w_i x_i)} - 1 \right]. \qquad (7.2)$$

7.3 Pareto tail modeling

The *Pareto* distribution is defined in terms of its cumulative distribution function

$$F_\theta(x) = 1 - \left(\frac{x}{x_0} \right)^{-\theta}, \qquad x \geq x_0, \qquad (7.3)$$

where $x_0 > 0$ is the scale parameter and $\theta > 0$ is the shape parameter (Kleiber and Kotz 2003). Furthermore, the density is given by

$$f_\theta(x) = \frac{\theta x_0^\theta}{x^{\theta+1}}, \qquad x \geq x_0. \qquad (7.4)$$

In Pareto tail modeling, the cumulative distribution function on the whole range of x is modeled as

$$F(x) = \begin{cases} G(x), & \text{if } x \leq x_0, \\ G(x_0) + (1 - G(x_0)) F_\theta(x), & \text{if } x > x_0, \end{cases} \qquad (7.5)$$

where G is an unknown distribution function (Dupuis and Victoria-Feser 2006).

Let n be the number of observations and let $\boldsymbol{x} = (x_1, \ldots, x_n)'$ denote the observed values with $x_1 \leq \ldots \leq x_n$. In addition, let k be the number of observations to be used for tail modeling. In this scenario, the threshold x_0 is estimated by

$$\hat{x}_0 := x_{n-k}. \qquad (7.6)$$

On the other hand, if an estimate $\hat{x}_0$ for the scale parameter of the Pareto distribution has been obtained, k is given by the number of observations larger than $\hat{x}_0$. Thus estimating x_0 and k directly corresponds with each other. Various methods for the estimation of x_0 or k have been proposed (Beirlant et al. 1996a,b, Dupuis and Victoria-Feser 2006, Van Kerm 2007). However, this paper is focused on evaluating robust methods for estimating the shape parameter θ (with respect to their influence on the selected Laeken indicators) once the threshold is fixed.

7.3.1 Hill estimator

The maximum likelihood estimator for the shape parameter of the Pareto distribution was introduced by Hill (1975) and is referred to as the *Hill* estimator. It is given by

$$\hat{\theta} = \frac{k}{\sum_{i=1}^{k} \log x_{n-k+i} - k \log x_{n-k}}. \qquad (7.7)$$

Note that the Hill estimator is non-robust, therefore it is included for benchmarking purposes.

7.3.2 Weighted maximum likelihood (WML) estimator

The weighted maximum likelihood (WML) estimator (Dupuis and Morgenthaler 2002, Dupuis and Victoria-Feser 2006) falls into the class of M-estimators and is given by the solution $\hat{\theta}$ of

$$\sum_{i=1}^{k} \Psi(x_{n-k+i}, \theta) = 0 \qquad (7.8)$$

with

$$\Psi(x, \theta) := w(x, \theta) \frac{\partial}{\partial \theta} \log f(x, \theta) = w(x, \theta) \left(\frac{1}{\theta} - \log \frac{x}{x_0} \right), \qquad (7.9)$$

where $w(x, \theta)$ is a weight function with values in $[0, 1]$. In this paper, a Huber type weight function is used, as proposed in Dupuis and Victoria-Feser (2006). Let the logarithms of the relative excesses be denoted by

$$y_i := \log \left(\frac{x_{n-k+i}}{x_{n-k}} \right), \qquad i = 1, \ldots, k. \qquad (7.10)$$

In the Pareto model, these can be predicted by

$$\hat{y}_i := -\frac{1}{\theta} \log \left(\frac{k+1-i}{k+1} \right), \qquad i = 1, \ldots, k. \qquad (7.11)$$

The variance of y_i is given by

$$\sigma_i^2 := \sum_{j=1}^{i} \frac{1}{\theta^2 (k-i+j)^2}, \qquad i = 1, \ldots, k. \qquad (7.12)$$

Using the standardized residuals

$$r_i := \frac{y_i - \hat{y}_i}{\sigma_i}, \qquad (7.13)$$

the Huber type weight function with tuning constant c is defined as

$$w(x_{n-k+i}, \theta) := \begin{cases} 1, & \text{if } |r_i| \leq c, \\ \frac{c}{|r_i|}, & \text{if } |r_i| > c. \end{cases} \qquad (7.14)$$

For this choice of weight function, the bias of $\hat{\theta}$ is approximated by

$$\hat{B}(\hat{\theta}) = -\frac{\sum_{i=1}^{k} \left(w_i \frac{\partial}{\partial \theta} \log f_i\right)|_{\hat{\theta}} \left(F_{\hat{\theta}}(x_{n-k+i}) - F_{\hat{\theta}}(x_{n-k+i-1})\right)}{\sum_{i=1}^{k} \left(\frac{\partial}{\partial \theta} w_i \frac{\partial}{\partial \theta} \log f_i + w_i \frac{\partial^2}{\partial \theta^2} \log f_i\right)|_{\hat{\theta}} \left(F_{\hat{\theta}}(x_{n-k+i}) - F_{\hat{\theta}}(x_{n-k+i-1})\right)}, \qquad (7.15)$$

where $w_i := w(x_{n-k+i}, \theta)$ and $f_i := f(x_{n-k+i}, \theta)$. This term is used to obtain a bias-corrected estimator

$$\tilde{\theta} := \hat{\theta} - \hat{B}(\hat{\theta}). \qquad (7.16)$$

For details and proofs of the above statements, the reader is referred to Dupuis and Morgenthaler (2002), Dupuis and Victoria-Feser (2006).

7.3.3 Partial density component (PDC) estimator

For the partial density component (PDC) estimator (Vandewalle et al. 2007), the Pareto distribution is modeled in terms of the relative excesses

$$y_i := \frac{x_{n-k+i}}{x_{n-k}}, \qquad i = 1, \ldots, k. \qquad (7.17)$$

The density function of the Pareto distribution for the relative excesses is approximated by

$$f_\theta(y) = \theta y^{-(1+\theta)}. \qquad (7.18)$$

The PDC estimator is then given by

$$\hat{\theta} = \arg\min_\theta \left[w^2 \int f_\theta^2(y) dy - \frac{2w}{k} \sum_{i=1}^{k} f_\theta(y_i)\right], \qquad (7.19)$$

i.e., by minimizing the integrated squared error criterion (Terrell 1990) using an incomplete density mixture model $w f_\theta$. The parameter w can be interpreted as a measure of the uncontaminated part of the sample and is estimated by

$$\hat{w} = \frac{\frac{1}{k} \sum_{i=1}^{k} f_{\hat{\theta}}(y_i)}{\int f_{\hat{\theta}}^2(y) dy}. \qquad (7.20)$$

See Vandewalle et al. (2007) and references therein for more information on the PDC estimator.

7.4 Simulation study

Various robust methods for the estimation of poverty and inequality indicators, mostly nonparametric, have been investigated in Van Kerm (2007), but neither the WML nor the PDC

7. A comparison of robust methods for Pareto tail modeling

estimator for Pareto tail modeling are considered there. Preliminary results with income generated from theoretical distributions (Holzer 2009) are an indication that both estimators are promising in the context of Laeken indicators. This is further investigated in this section. However, variance estimation is not yet considered in this paper.

The simulations are carried out in R (R Development Core Team 2010) using the package **simFrame** (Alfons 2010, Alfons et al. 2010c), which is a general framework for statistical simulation studies. A synthetic data set consisting of 35 041 households and 81 814 individuals is used as population data in the simulation study. This data set has been generated with the R package **simPopulation** (Alfons et al. 2010b, Alfons and Kraft 2010) based on Austrian EU-SILC survey data from 2006 and is about 1% of the size of the real Austrian population. A thorough investigation in a close-to-reality environment using real-life sized synthetic Austrian population data is future work.

From the synthetic data, 500 samples are drawn using simple random sampling. Each sample consists of 6 000 households, which is roughly the sample size used in the real-life survey. With these samples, two scenarios are investigated. In the first scenario, no contamination is added. In the second, the equivalized disposable income of 0.25% of the households is contaminated. The contamination is thereby drawn from a normal distribution with mean $\mu = 1\,000\,000$ and standard deviation $\sigma = 10\,000$. Note that the *cluster effect* is considered, i.e., all persons in a contaminated household receive the same income. The threshold for Pareto tail modeling is in both cases set to $k = 275$ based on graphical exploration of the original EU-SILC sample with a Pareto quantile plot (Beirlant et al. 1996a). Furthermore, the tuning constant $c = 2.5$ is used for the bias-corrected WML estimator due to favorable robustness properties (Holzer 2009).

Figure 7.1 shows the results of the simulations without contamination for the quintile share ratio *(left)* and the Gini coefficient *(right)*. The three methods for tail modeling as well as the standard estimation method without tail modeling behave very similarly and are very close to the true values, which are represented by the grey lines. This is also an indication that the choice of k is suitable.

Figure 7.2 shows the results of the simulations with 0.25% contamination for the quintile share ratio *(left)* and the Gini coefficient *(right)*. Even such a small amount of contamination completely corrupts the standard estimation of these inequality indicators. Fitting the Pareto distribution with the Hill estimator is still highly influenced by the outliers. The best results are obtained with the PDC estimator, while the WML estimator shows a slightly larger bias.

7.5 Conclusions and outlook

The quintile share ratio and the Gini coefficient, which are inequality indicators belonging to the set of Laeken indicators, are highly influenced by outliers. A simulation study for

7.5 Conclusions and outlook

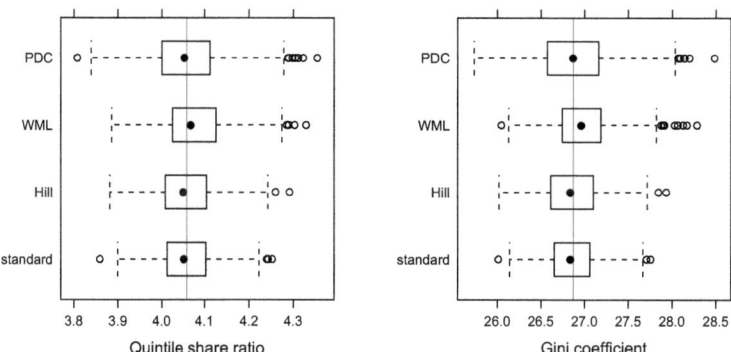

Figure 7.1: Simulation results for the quintile share ratio *(left)* and the Gini coefficient *(right)* without contamination.

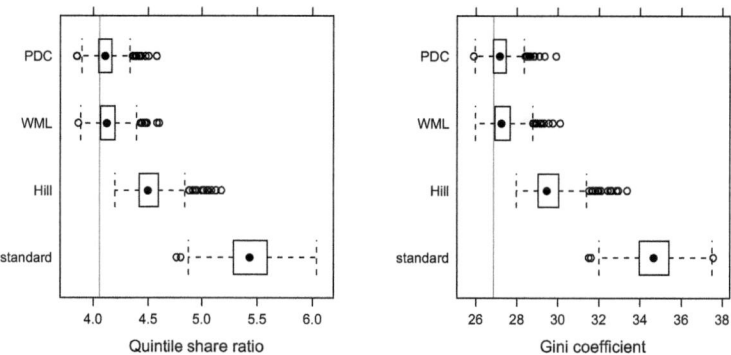

Figure 7.2: Simulation results for the quintile share ratio *(left)* and the Gini coefficient *(right)* with 0.25% contamination.

the case of simple random sampling showed that robust Pareto tail modeling can be used to reduce the influence of the outlying observations. The partial density component (PDC) estimator thereby performed best.

The simulation study in this paper is limited to simple random sampling because the estimators for Pareto tail modeling do not account for sample weights. Future work is to modify the estimators such that sample weights are taken into account, to investigate variance estimation, and to perform simulations using real-life sized synthetic population data.

Acknowledgements This work was partly funded by the European Union (represented by the European Commission) within the 7th framework programme for research (Theme 8, Socio-Economic Sciences and Humanities, Project AMELI (Advanced Methodology for European Laeken Indicators), Grant Agreement No. 217322). Visit `http://ameli.surveystatistics.net` for more information on the project.

Chapter 8

Robust variable selection with application to quality of life research[1]

To appear in the journal *Statistical Methods & Applications* (Alfons et al. 2011a).

Andreas Alfons[a], Wolfgang E. Baaske[b], Peter Filzmoser[a], Wolfgang Mader[c], Roland Wieser[b]

[a] Department of Statistics and Probability Theory, Vienna University of Technology
[b] STUDIA-Schlierbach, Studienzentrum für internationale Analysen
[c] SPES Academy

Abstract A large database containing socioeconomic data from 60 communities in Austria and Germany has been built, stemming from 18 000 citizens' responses to a survey, together with data from official statistical institutes about these communities. This paper describes a procedure for extracting a small set of explanatory variables to explain response variables such as the cognition of quality of life. For better interpretability, the set of explanatory variables needs to be very small and the dependencies among the selected variables need to be low. Due to possible inhomogeneities within the data set, it is further required that the solution is robust to outliers and deviating points. In order to achieve these goals, a robust model selection method, combined with a strategy to reduce the number of selected predictor variables to a necessary minimum, is developed. In addition, this context-sensitive method is applied to obtain responsible factors describing quality of life in communities.

Keywords Robustness, Model selection, Success factors, Quality of life

---

[1] The research was supported by a grant of the Austrian Research Promotion Agency (FFG), Project Ref. No. 813000/10345.

8.1 Introduction

The research project *ErfolgsVision* (English: *vision of success*) is a joint cooperation of the Austrian institutions SPES Academy (a regional developer), STUDIA-Schlierbach (an applied social researcher) and the Department of Statistics and Probability Theory at Vienna University of Technology. For this project, data from screening processes carried out by SPES in 60 communities in Austria and Germany during the period of 2000 to 2006 were used. In total, 18 748 questionnaires were collected, on average 312 per municipality. The survey was subject to individual adaptations towards the needs of the municipalities. It usually comprised about 250 questions, most of them multiple choice. In this project, we were interested in comparing the communities, therefore indicators referring to the questions were computed jointly from the questionnaires of each community. These data were merged with statistics on demography and economy. After removing observations with more than 50% and variables with more than 20% of missing values, a data matrix with 43 (out of 60) observations and 153 (out of 250) variables resulted. Some of the observations still included missing values (in one case for 20% of the variables), thus kNN imputation (Troyanskaya et al. 2001) was used to obtain a complete data matrix.

Although the goal of the project was much broader, this paper is focused on finding the factors controlling quality of life. Since an easy interpretation of the results was a major objective, the number of explanatory variables should be limited to about 5 to at most 10. Moreover, the analysis needed to be robust against outliers and deviating data points because of possible inhomogeneities within the data set.

Various methods for model selection have been proposed to date. Here we are interested in robust approaches, as they are less sensitive to outliers in the data. Such methods have gained increasing attention in the literature (e.g., Ronchetti and Staudte 1994, Ronchetti et al. 1997, Wisnowski et al. 2003, Müller and Welsh 2005, Khan et al. 2007a,b, McCann and Welsch 2007, Salibian-Barrera and Van Aelst 2008, Choi and Kiefer 2010, Riani and Atkinson 2010, Van Aelst et al. 2010). However, robust variable selection is especially difficult if the number of observations is smaller than the number of variables. In that case it is no longer possible to directly apply robust regression methods (Maronna et al. 2006) in order to select the most significant variables. On the other hand, various techniques for variable selection in high dimensions have been introduced, which are based on the non-robust least squares criterion (see, e.g., Hastie et al. 2009, Varmuza and Filzmoser 2009). An example is least angle regression (LARS; Efron et al. 2004), which selects the regressor variables in the order of their importance for predicting the response variable. LARS has been robustified in Khan et al. (2007b) by two different approaches: the *plug-in* method and the *cleaning* method. In the plug-in method, the non-robust estimators mean, variance and correlation in classical LARS are replaced by robust counterparts. The idea of the cleaning method, on the other hand, is to shrink outliers and to apply classical LARS to the cleaned data. Both methods use the so-called *winsorization* technique to estimate the correlations and shrink the outliers, respectively. Thus the influence of potential outliers on computing the sequence of predictors is reduced. Since the plug-in approach is computationally faster and more widely applicable, it is the basis of our algorithm for robust variable selection. In the following, the plug-in method will be referred to as RLARS. Khan et al. (2007b) illustrated that the sequence of predictors returned by RLARS can be stabilized with the help of the bootstrap. The resulting procedure is called *bootstrapped* RLARS, for short B-RLARS.

A reduced set of the B-RLARS sequence of candidate predictors is then used for building a more refined regression model. For this purpose we suggest to use MM-regression (Yohai 1987, Maronna et al. 2006). MM-estimators have many desirable properties. Most importantly, they combine a maximum breakdown point of 0.5 with high efficiency. Salibian-Barrera and Zamar (2002) further studied the distribution of MM-estimates using a robust bootstrap method. We apply MM-regression to filter out the non-significant variables at a certain significance level. Since in general the resulting number of the resulting variables is still too high for a reasonable interpretation, all possible subsets of size k are examined (see, e.g., Furnival and Wilson 1974, Miller 2002, Gatu and Kontoghiorghes 2006), which is sometimes referred to as k-subset regression. In our case, a robustified version of k-subset regression is applied by using the weights obtained from MM-regression. Thus strong dependencies among the regressor variables are eliminated and the smaller models are highly interpretable, which is required in the context of social sciences. This approach will therefore be called *context-sensitive* and can be considered a trade-off between quality of the model and interpretability.

The rest of this paper is organized as follows. In Section 8.2, we will describe the complete algorithm in more detail. Section 8.3 outlines how the procedure can be applied to obtain a small set of explanatory variables determining quality of life, and a simulation study is performed in Section 8.4. The final Section 8.5 concludes.

8.2 Context-sensitive model selection

Let $y = (y_1, \ldots, y_n)^t$ be the response variable and $x_1 = (x_{11}, \ldots, x_{n1})^t$, ..., $x_p = (x_{1p}, \ldots, x_{np})^t$ the candidate predictors. Thus n denotes the number of observations and p the number of candidate predictors. Furthermore, let $J = \{1, \ldots, p\}$ be the set of indices referring to the candidate predictor variables. Our method aims to find a model for the response variable y that contains a very low number of predictors, at most $k \ll p$, in order to achieve high interpretability. Since the predictor variables should contain potentially new information, an additional requirement is that strong dependencies among the regressor variables should be avoided. These goals of easy-to-interpret models and low or only moderate dependencies between the predictors reflect the context-sensitivity of our method.

8.2.1 Description of the algorithm

For a start, the response variable y and the candidate predictors $x_1, \ldots, x_p$ are robustly centered and scaled using median and MAD, according to

$$y_i^* = \frac{y_i - med(y_1, \ldots, y_n)}{MAD(y_1, \ldots, y_n)}, \qquad i = 1, \ldots, n \qquad (8.1)$$

$$x_{ij}^* = \frac{x_{ij} - med(x_{1j}, \ldots, x_{nj})}{MAD(x_{1j}, \ldots, x_{nj})}, \qquad i = 1, \ldots, n, \; j = 1, \ldots, p. \qquad (8.2)$$

Hence all predictor variables $x_j^* = (x_{1j}^*, \ldots, x_{nj}^*)^t$, $j = 1, \ldots, p$, are on an equal scale. Our algorithm then proceeds in three steps. The first step seeks a drastic reduction of the number of candidate predictors such that the following steps become computationally feasible. For this purpose, B-RLARS (Khan et al. 2007b) is applied to $y^* = (y_1^*, \ldots, y_n^*)^t$ and $x_1^*, \ldots, x_p^*$ to find a sequence $(x_j^*)_{j \in J_1}$, $J_1 \subset J$, of candidate predictors for y^* with $k < |J_1| \ll p$. Clearly, J_1 contains the indices of the $|J_1|$ most important predictor variables returned by B-LARS. In order to allow for an interpretation of the final model, $|J_1|$ should be in the range of 10 to 20.

In the second step, the covariates x_j^*, $j \in J_1$, are entered as predictors for y^* in MM-regression (Yohai 1987, Maronna et al. 2006). We apply MM-regression to filter out the non-significant variables. Let $J_2 \subseteq J_1$ be the set of indices of the significant variables at a given significance level α. The choice of α should not be too strict (we used $\alpha = 0.3$) in order not to exclude important variables. Note that this test is robust because it is based on robust estimates of the standard errors (Croux et al. 2008). The second step thus concludes with fitting another MM-regression model to y^*, using only the significant predictors x_j^*, $j \in J_2$. Thus we consider the regression model

$$y_i^* = (\mathbf{x}_i^*)^t \boldsymbol{\beta} + e_i, \qquad i = 1, \ldots, n, \qquad (8.3)$$

where $\mathbf{x}_i^*$ denotes the i-th observation of the predictor variables x_j^*, $j \in J_2$, extended by 1 in the first component to account for the intercept. Furthermore, $\boldsymbol{\beta}$ is the vector of length $|J_2| + 1$ of the unknown regression coefficients, and e_i denotes the error terms, which are assumed to be i.i.d. random variables. MM-regression minimizes a function of the scaled residuals. Denoting the residuals by $r_i(\boldsymbol{\beta}) = y_i^* - (\mathbf{x}_i^*)^t \boldsymbol{\beta}$, MM-regression solves the problem

$$\hat{\boldsymbol{\beta}} = \operatorname*{argmin}_{\boldsymbol{\beta}} \sum_{i=1}^{n} \rho\left(\frac{r_i(\boldsymbol{\beta})}{\hat{\sigma}}\right), \qquad (8.4)$$

where $\rho(r)$ is a bounded function, and $\hat{\sigma}$ is a robust scale estimator of the residuals, derived from a robust (but inefficient) S-estimator (for more details, see Maronna et al. 2006). Differentiating (8.4) with respect to $\boldsymbol{\beta}$ yields

$$\sum_{i=1}^{n} \psi\left(\frac{r_i(\boldsymbol{\beta})}{\hat{\sigma}}\right) \mathbf{x}_i^* = 0 \quad (8.5)$$

where $\psi = \rho'$. Using the notation

$$w_i = \frac{\psi(r_i(\boldsymbol{\beta})/\hat{\sigma})}{r_i(\boldsymbol{\beta})/\hat{\sigma}}, \quad i = 1, ..., n, \quad (8.6)$$

allows (8.5) to be rewritten as

$$\sum_{i=1}^{n} w_i r_i(\boldsymbol{\beta}) \mathbf{x}_i^* = 0. \quad (8.7)$$

Equation (8.7) is a weighted version of the normal equations. Hence the estimator can be considered a weighted least squares estimator with weights w_i from (8.6), which depend on the data. For an estimator to be robust, observations with large residuals should receive small weights. Thus the function ρ was chosen as the bisquare function (see Maronna et al. 2006), which ensures that $\psi(r)$ is decreasing towards zero for increasing $|r|$. The resulting weights $\hat{w}_i$, $i = 1, \ldots, n$, for the MM-regression estimator $\hat{\boldsymbol{\beta}}$ will be used in the third step of the algorithm.

The third step is based on k-subset regression (see, e.g., Furnival and Wilson 1974, Miller 2002, Gatu and Kontoghiorghes 2006). Thus we want to find the best subset of maximum size k of the predictor variables that optimizes a criterion such as Mallows' C_p (Mallows 1973) or the BIC (Schwarz 1978). Although k-subset regression is not feasible even for moderate numbers of predictors, our method does not suffer from this problem since the number of predictors has been drastically reduced with B-RLARS in the first step and MM-regression in the second step. Another problem with k-subset regression is that it is not robust. However, a simple robustification is to use the weights computed in the second step during MM-regression, i.e., to enter the procedure with the response variable $\tilde{\boldsymbol{y}} = (\hat{w}_1 y_1^*, \ldots, \hat{w}_n y_n^*)^t$ and the candidate predictors $\tilde{\boldsymbol{x}}_j = (\hat{w}_1 x_{1j}^*, \ldots, \hat{w}_n x_{nj}^*)^t$, $j \in J_2$. Since the data are robustly standardized, multiplying the observations with the weights results in shrinking the outliers towards the main body of the data. This robustified version of k-subset regression yields the optimal subset $\{\boldsymbol{x}_j^* : j \in J_3\}$ with $J_3 \subseteq J_2, |J_3| \leq k$, of the set of candidate predictors $\{\boldsymbol{x}_j^* : j \in J_2\}$.

Instead of using the weights computed in the second step, other robust versions of k-subset regression might be considered. One example is fitting MM-regression models to all possible subsets of maximum size k and using m-fold cross-validation to estimate a robust prediction loss function, e.g., the root trimmed mean squared error of prediction (RTMSEP),

8. Robust variable selection

for choosing the optimal submodel. In m-fold cross validation, the data are split randomly in m blocks of approximately equal size. Each block is left out once for fitting the model, and the left-out block is used as test data. Thus a prediction is obtained for each observation. Let $b(i)$ be the block to which observation $i = 1, \ldots, n$ belongs, then the prediction for y_i is denoted by $\hat{y}_i^{-b(i)}$. For a trimming factor $0 \leq \gamma < 0.5$, the RTMSEP is defined as

$$\text{RTMSEP} = \sqrt{\frac{1}{N} \sum_{i=1}^{N} r_{(i)}^2} \qquad (8.8)$$

where $r_i = y_i - \hat{y}_i^{-b(i)}$, $i = 1, \ldots, n$, are the residuals using the predictions from cross-validation, $r_{(1)}^2 \leq \ldots \leq r_{(n)}^2$ are the sorted squared residuals, and $N = n - \lfloor n\gamma \rfloor$ (here $\lfloor a \rfloor$ denotes the integer part of a). Whereas such procedures are certainly more robust than the simple weighted approach, they are computationally expensive even for small problems. On the other hand, using the weights computed in the second step of the procedure results in a cleaned data set, thus reducing the influence of atypical observations in both fitting the submodels and computing classical criteria for deciding on the best submodel. Even though the weights might not be optimal for each submodel, this approach is a reasonable compromise between computational complexity and robustness. It is fast for small problems and worked very well in our studies (see the example in Section 8.3).

8.2.2 Summary of the algorithm

The response variable and all candidate predictor variables are robustly centered and scaled using median and MAD. The resulting response variable is denoted by $\boldsymbol{y}^* = (y_1^*, \ldots, y_n^*)^t$, and the resulting candidate predictors by $\boldsymbol{x}_1^* = (x_{11}^*, \ldots, x_{n1}^*)^t, \ldots, \boldsymbol{x}_p^* = (x_{1p}^*, \ldots, x_{np}^*)^t$. Let $J = \{1, \ldots, p\}$ be the set of indices for the candidate predictors, and $k \ll p$ the desired maximum number of predictors for the model. Then the algorithm can be summarized as follows:

1. Perform B-RLARS on $\boldsymbol{y}^*$ and $\boldsymbol{x}_1^*, \ldots, \boldsymbol{x}_p^*$ to compute a sequence $(\boldsymbol{x}_j^*)_{j \in J_1}$, $J_1 \subset J$, of candidate predictors with $k < |J_1| \ll p$.

2. Use $\boldsymbol{x}_j^*$, $j \in J_1$, as predictors for $\boldsymbol{y}^*$ in MM-regression. Let $J_2 \subseteq J_1$ be the set of indices of the significant variables at a given significance level α. Fit another MM-regression model to $\boldsymbol{y}^*$ with only the significant predictors $\boldsymbol{x}_j^*$, $j \in J_2$, and let $\hat{w}_1, \ldots, \hat{w}_n$ denote the resulting weights for the observations.

3. Apply k-subset regression with the response variable $\tilde{\boldsymbol{y}} = (\hat{w}_1 y_1^*, \ldots, \hat{w}_n y_n^*)^t$ and the candidate predictors $\tilde{\boldsymbol{x}}_j = (\hat{w}_1 x_{1j}^*, \ldots, \hat{w}_n x_{nj}^*)^t$, $j \in J_2$. This robustified version of k-

subset regression yields the optimal subset $\{\boldsymbol{x}_j^* : j \in J_3\}$ with $J_3 \subseteq J_2, |J_3| \leq k$, of the set of candidate predictors $\{\boldsymbol{x}_j^* : j \in J_2\}$.

A more visual summary of the algorithm is given by the following diagram:

$$
\begin{array}{ccccccc}
& \text{B-RLARS} & & \text{MM-regression} & & k\text{-subset regression} & \\
J & \longrightarrow & J_1 & \longrightarrow & J_2 & \longrightarrow & J_3
\end{array}
$$

8.2.3 Diagnostics

The elimination of high dependencies among the predictor variables is a major demand for our context-sensitive method. In the social sciences, such a model has potential for an interesting interpretation. Correlated predictor variables, on the other hand, are likely to describe more or less the same factors, which are just expressed with different variables in the data set. The resulting model will not be as interesting with respect to interpretation, even if it has a high prediction ability of the response variable. Hence a graphical tool to check whether the procedure succeeded in fulfilling this demand would be useful. A dendrogram (e.g., Everitt and Dunn 2001) based on robust correlations seems suitable for this purpose.

Since the number of candidate predictors is in general too large for an informative plot, only the variables $\boldsymbol{x}_j$, $j \in J_1$, from the initial B-RLARS sequence will be used. The correlation matrix of this reduced set of candidate predictors can be estimated with a high-breakdown estimator such as the *minimum covariance determinant* (MCD; Rousseeuw and Van Driessen 1999) or the *orthogonalized Gnanadesikan-Kettenring* estimator (OGK; Maronna and Zamar 2002). Note that the correlations used here do not need not come from an affine equivariant or orthogonal equivariant method, the Spearmann or Kendall correlation could also be used (for their robustness properties, see Croux and Dehon 2010). Let $\boldsymbol{R} = (r_{ij})_{i,j \in J_1}$ denote such a robust estimate of the correlation matrix. Then the dissimilarity matrix $\boldsymbol{D} = (d_{ij})_{i,j \in J_1}$ given by

$$d_{ij} = 1 - |r_{ij}|, \qquad i, j \in J_1, \tag{8.9}$$

is used for clustering the variables. *Complete linkage* clustering (e.g., Everitt and Dunn 2001) is well suited for our purposes, as the dissimilarity measure is based on robust correlations. In this method, the dissimilarity of two clusters A and B is defined as

$$d(A, B) = \max_{\boldsymbol{x}_i \in A, \boldsymbol{x}_j \in B} d_{ij}. \tag{8.10}$$

Using (8.9), this can be written as

$$d(A, B) = 1 - \min_{\boldsymbol{x}_i \in A, \boldsymbol{x}_j \in B} |r_{ij}|. \tag{8.11}$$

8. Robust variable selection

In each step, the two clusters with minimum dissimilarity are merged. Thus complete linkage clustering in our case yields that variables with low correlations will not belong to the same cluster if an appropriate cut-off point is chosen. Hence the resulting dendrogram is a convenient way of exploring the robust correlation structures among the candidate predictor variables. If the selected variables belong to different clusters, then the procedure performed well in the context-sensitive sense. Such a dendrogram may also reveal potential problems due to strong correlations among all predictor variables. In this case, it would probably be difficult to decide on which variables should be eliminated for a highly interpretable model.

8.2.4 Implementation

An implementation of our algorithm in the statistical environment R (R Development Core Team 2010) and detailed documentation can be downloaded from http://www.statistik.tuwien.ac.at/public/filz/programs.html. The required R code for B-RLARS by Khan et al. (2007b) can be obtained from http://users.ugent.be/~svaelst/software/RLARS.html. In addition, the R packages **robustbase** (Rousseeuw et al. 2009) and **leaps** (Lumley and Miller 2009), which are available on CRAN (the Comprehensive R Archive Network, http://CRAN.R-project.org), need to be installed.

8.3 Example: Driving factors behind quality of life

In this section, we will attempt to find the driving factors behind quality of life in communities, using the data collected by SPES (see Section 8.1 for a general description of the data). Table 8.1 contains explanations for the most important variables. In order to ensure an easy-to-interpret model, the response variable *qualityLife* should be explained by at most 10 predictors. Note that some variables, which are too discontinuous or clearly redundant in the context of quality of life, are removed from the data set, resulting in 138 remaining candidate predictors. Hence all variables are continuous, which is important for applying the developed robust method.

Furthermore, we will compare our robust context-sensitive method, in the following referred to as RCS, with B-RLARS.

8.3.1 Results

RCS is carried out with parameter settings as described in the following. As mentioned above, the maximum number of variables in the final model is set to $k = 10$. In the initial B-RLARS step, 15 variables are sequenced with 50 bootstrap repetitions. These candidate predictors are then filtered at significance level $\alpha = 0.3$ in MM-regression. This unusually

Table 8.1: Explanation of important variables.

Variable	Explanation
qualityLife	quality of life
agriculture	state of local agriculture
beauty	beauty of the community
contrFarmers	contribution of local farmers to quality of life
futureComm	future development of the community
impOrganic	importance of organic products
impTrad	importance of traditional festivities
interesting	interestingness of the community
medCare	state of medical care
merchAssort	assortment of local merchants
merchComm	contribution of local merchants to the development of the community
parish	state of local parish
percAdolesc	percentage of adolescents
publicServ	state of public services
eduProTraining	educational and professional training opportunities
view	state of the community's view

high significance level will prevent the exclusion of potentially important variables. For deciding on the optimal submodel in the robustified version of k-subset regression, the BIC is used as criterion. With these parameters, RCS returns the following six predictors: *agriculture*, *medCare*, *merchAssort*, *eduProTraining*, *beauty* and *parish* (see Table 8.1).

In addition to the simple weighted k-subset regression in the third step of RCS, we also apply a more sophisticated robust version for comparison. In this version, we fit MM-regression models to the subsets and use fivefold cross-validation to estimate the root trimmed mean squared error of prediction (RTMSEP) with 20% trimming, see (8.8). Fivefold cross-validation seems to be a reasonable choice given the number of observations in the data set. Furthermore, the choice of the trimming proportion is based on the weights returned by the MM-regression in the second step, which indicate some outliers. With a lower value, these outliers may still influence the RTMSEP, whereas a higher value may result in some bias. The submodel with the lowest RTMSEP is then chosen as the optimal submodel. While this procedure yields the same six variables as the simple weighted approach, it is computationally much more expensive.

In order to compare RCS with B-RLARS, we start with the B-RLARS sequence of length 15 that we computed in the first step of RCS. Then we proceed as in the examples in Section 6 of Khan et al. (2007b) to obtain the final B-RLARS model. There it is suggested to start with the first variable and to increase the number of variables along the sequence, while

Table 8.2: MM-regression results for the RCS model for quality of life.

	Estimate	Standard error	t-Value	p-Value
(Intercept)	-2.302	11.227	-0.205	0.839
agriculture	0.251	0.053	4.713	$3.6 \cdot 10^{-5}$
medCare	0.076	0.023	3.228	0.003
merchAssort	0.177	0.064	2.751	0.009
eduProTraining	0.117	0.026	4.450	$8.0 \cdot 10^{-5}$
beauty	0.292	0.113	2.588	0.014
parish	0.216	0.035	6.226	$3.5 \cdot 10^{-7}$

Robust residual standard error: 1.705

Table 8.3: MM-regression results for the B-RLARS model for quality of life.

	Estimate	Standard error	t-Value	p-Value
(Intercept)	8.795	7.079	1.242	0.221
agriculture	0.337	0.064	5.278	$5.2 \cdot 10^{-6}$
merchAssort	0.277	0.065	4.297	$1.1 \cdot 10^{-4}$
interesting	0.409	0.082	5.009	$1.2 \cdot 10^{-5}$

Robust residual standard error: 2.419

fitting a robust regression model in each step. For each model, the robust R^2 measure

$$R_{rob}^2 = 1 - \left(\frac{med(|y_1 - \hat{y}_1|, \ldots, |y_n - \hat{y}_n|)}{MAD(y_1, \ldots, y_n)} \right)^2, \qquad (8.12)$$

is computed, where y_i, $i = 1, \ldots, n$, are the observed values of the response variable and $\hat{y}_i$, $i = 1, \ldots, n$, are the fitted values (see Rousseeuw and Leroy 1987). Finally, these robust R^2 values are plotted against the model size to obtain a *learning curve* (c.f. Croux et al. 2003). Note that the robust R^2 is not always monotonically increasing with the number of variables since algorithms for robust regression yield only approximate solutions. Keeping in mind that the number of predictors should be at most 10, the learning curve in Figure 8.1 (*left*) suggests using the first 8 variables of the sequence: *contrFarmers, agriculture, medCare, merchComm, impOrganic, merchAssort, percAdolesc* and *interesting* (see Table 8.1). These variables are further examined by fitting MM-regression models to all possible subsets. Deciding on the best subset is done by minimizing the RTMSEP with 20% trimming, which is estimated using fivefold cross-validation. The final model resulting from this procedure contains the predictors *agriculture, merchAssort* and *interesting*.

Tables 8.2 and 8.3 show the results of MM-regression with the predictor variables selected by RCS and B-RLARS, respectively. In both models, the included variables are highly

8.3 Example: Driving factors behind quality of life

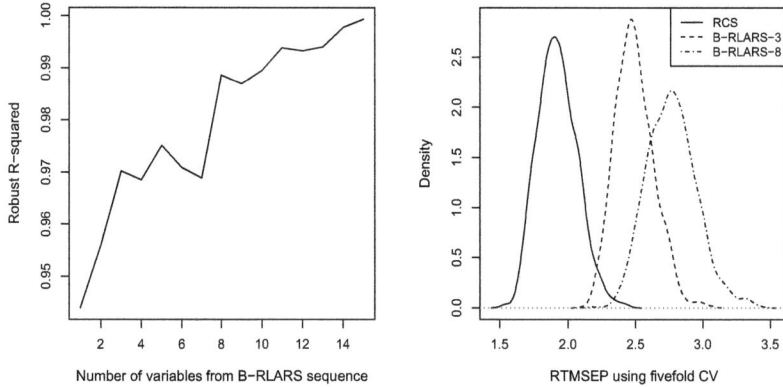

Figure 8.1: Learning curve for the B-RLARS sequence (*left*). Densities of the RTMSEP for the RCS model, the final B-RLARS model (B-RLARS-3) and the B-RLARS model with the first 8 variables as suggested by the learning curve (B-RLARS-8), estimated with repeated fivefold cross-validation (*right*).

significant. Containing only three predictor variables, the B-RLARS model is on the one hand somewhat simpler than the RCS model, which consists of six predictors. Two of the three variables selected by B-RLARS are also selected by RCS (*agriculture* and *merchAssort*). On the other hand, the robust residual standard error indicates that the B-RLARS model might be too simple. The RCS model is a better fit due to the much lower robust residual standard error.

However, in order to decide on which model is preferable, it is necessary to estimate the prediction quality of the models. For this purpose, repeated fivefold cross-validation with 1,000 repetitions is applied. In each repetition, the RTMSEP with 20% trimming is estimated. Figure 8.1 (*right*) displays the resulting density curves for the RCS model, the final B-RLARS model (B-RLARS-3) and the B-RLARS model with the first 8 variables as suggested by the learning curve (B-RLARS-8). It is clearly visible from this plot that the average RTMSEP is significantly smaller for RCS than for the other two models. Even though the variance of the RTMSEP is slightly larger for RCS than for B-RLARS-3, it is comparable for the two methods. Thus the RCS model performs much better than the two B-RLARS models, while the B-RLARS-8 model clearly leads to the worst prediction performance.

One of the main requirements concerning context-sensitivity was that the resulting model should be simple. Nevertheless, while succeeding in finding a few important predictor variables, the B-RLARS model turns out to be too simple. By only moving along the computed

8. Robust variable selection

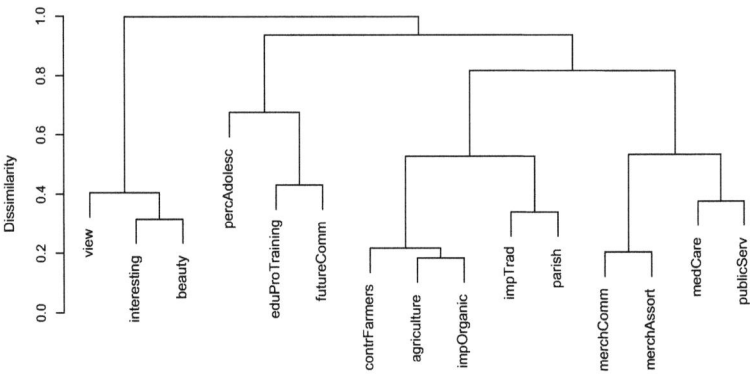

Figure 8.2: Dendrogram (based on robust correlations) of the initial B-RLARS sequence of candidate predictors for quality of life.

sequence of candidate predictors for finding the optimal size of the model, variables such as *medCare* and *eduProTraining* were completely neglected, even though they are clearly very important in the context of quality of life. Since RCS manages to include these variables in the selected model, the key step for context-sensitivity in the RCS procedure may be selecting the variables of the initial B-RLARS sequence at a certain significance level in MM-regression.

Another main requirement was that the dependencies among the selected variables should be rather low. Therefore, a dendrogram is constructed according to Section 8.2.3 and shown in Figure 8.2. It includes the 15 most important candidate predictors for quality of life, which were sequenced with B-RLARS in the first step of our context-sensitive procedure. The robust correlations for the dendrogram were computed with the reweighted MCD. The trimming parameter for the size of the subsets was thereby set to 75%. Furthermore, the finite sample correction factor and the asymptotic consistency factor were used. The dendrogram shows that RCS was able to fulfill this demand of low variable dependencies. In addition, every group in the dendrogram is represented in the RCS model, but not in the B-RLARS model.

The results seem to be significant in terms of theoretical concepts for quality of life assessments. Our selection procedure definitely moves beyond producing inconsistent lists of indicators, it creates a set of meaningful empirical measures. In quality of life research (e.g., Diener et al. 1999), individualistic and subjective indicators prevail, but recent concepts combine them with features of the external world. The model of Renwick et al. (1994), followed by Tichbon and Newton (2002), allows subjective states (*being*—e.g., health, nutrition, beliefs, values), as well as objective states (*belonging*—e.g., services, activities, leisure) and development (*becoming*—e.g., acquisition of skills and knowledge). Meaningful variables of all three types are included in the empirical results presented in this article (with some of the variables loading on different types): *medCare* and *merchAssort* are *being*-indicators, *agriculture* and *beauty* are *belonging*-indicators, while *parish*, *interesting* and *eduProTraining* signify development (*becoming*). The studies of this project are insofar unique, as they combine internal and external world features on a solid data base with appropriate analysis techniques. We recommend to incorporate the results into the design of agricultural policies. Municipalities often underestimate the role of the "lagging-behind" sector agriculture, whereas our analysis shows that the state of local agriculture constitutes a significant share of quality of life. On a world-wide level, producing quality of life as an external effect within the proximity may cause agriculture to be be respected and handled differently from a mere producer of tradable commodities (Baaske et al. 2009).

8.3.2 CPU times

The computation times presented in this section are average times over 50 runs, carried out on a machine with an Intel Core2 Quad 2.66GHz processor and 8GB main memory. Keep in mind that the computations were carried out with R (and thus only one of the four available processors was effectively used), and that the data set consists of 43 observations and 138 candidate predictor variables. With the parameter settings as described in the beginning of Section 8.3.1, RCS completed after 20.61 seconds. The running time was thereby dominated by computing the initial B-RLARS sequence, which took 20.54 seconds. This example indicates that RCS is still feasible whenever computing the initial B-RLARS sequence is feasible.

For finding the final B-RLARS model, the learning curve had to be inspected graphically to find the optimal number of predictors. Afterwards, all subsets of the reduced sequence were examined using MM-regression and fivefold cross-validation, which was very time-consuming for such a small problem. Since RCS uses the simple weighted version of robust k-subset regression and does not require manual interaction, obtaining the RCS model was much faster than obtaining the final B-RLARS model.

8.4 Simulations

For further investigation of the proposed RCS procedure, simulations are carried out using a simulation setting similar to that from Khan et al. (2007b). With k latent independent standard normal variables $l_1, \ldots, l_k$ and an independent standard normal variable e, a linear

8. Robust variable selection

model is constructed as
$$y := l_1 + \ldots + l_k + \sigma e, \tag{8.13}$$
where σ is chosen so that the signal-to-noise ratio is 5, i.e.,
$$\sqrt{\text{var}(l_1 + \ldots + l_k)/\text{var}(\sigma e)} = \sqrt{k}/\sigma = 5. \tag{8.14}$$
Using independent standard normal variables $e_1, \ldots, e_p$, a set of p candidate predictors is then constructed as
$$\begin{aligned}
x_1 &:= l_1 + \tau e_1, \\
x_2 &:= l_1 + \tau e_2, \\
x_3 &:= l_1 + \tau e_3, \\
&\vdots \\
x_{3k-2} &:= l_k + \tau e_{3k-2}, \\
x_{3k-1} &:= l_k + \tau e_{3k-1}, \\
x_{3k} &:= l_k + \tau e_{3k}, \\
x_{3k+1} &:= l_1 + \delta e_{3k+1}, \\
x_{3k+2} &:= l_1 + \delta e_{3k+2}, \\
&\vdots \\
x_{5k-1} &:= l_k + \delta e_{5k-1}, \\
x_{5k} &:= l_k + \delta e_{5k}, \\
x_i &:= e_i, \quad i = 5k+1, \ldots, p,
\end{aligned} \tag{8.15}$$
where $\tau = 0.2$ and $\delta = 5$ so that $x_1, \ldots, x_{3k}$ form k groups of low-noise perturbations of the latent variables, $x_{3k+1}, \ldots, x_{5k}$ are noise covariates that are correlated with the latent variables, and $x_{5k+1}, \ldots, x_p$ are independent noise covariates.

Regarding contamination, the following scenarios are investigated (similar to a subset of the scenarios investigated in Khan et al. 2007b), where ε denotes the fraction of outliers in the data:

1. No contamination.

2. Contamination in y given by $e \sim (1-\varepsilon)N(0,1) + \varepsilon N(0,1)/U(0,1)$.

3. Same as 2., but contaminated observations contain outliers in $x_1, \ldots, x_p$ coming from $N(5,1)$.

Note that in the last scenario, the contamination is not more extreme because the outliers in the data for which the proposed method has been designed (see Section 8.1) are moderate as well.

In the simulation experiments in Khan et al. (2007b), B-RLARS is compared to other methods using *recall curves*, i.e., the average numbers of target variables included in the first m sequenced variables are plotted, with m varying within a certain range. However, our procedure does not produce a sequence of predictor variables, instead it is designed to obtain a final model from an initial sequence of candidate predictors. Hence a comparison with B-RLARS using recall curves is not meaningful.

Moreover, one requirement for our procedure is that strong correlations between variables should be avoided. For each latent variable, a group of low-noise perturbations is thus defined in (8.15). Variables in the same group are highly correlated, while the correlations between variables from different groups are low. The procedure is successful in the context-sensitive sense if the final model contains exactly one predictor variable from each of these groups. Nevertheless, the success of the procedure of course also depends on the initial B-RLARS sequence. If no variables of one group exist in the initial sequence, the final model cannot contain a variable of this group either.

In the simulations, $k = 5$ latent variables are used to construct the linear model for the response as in (8.13) and $p = 100$ candidate predictors as in (8.15). Concerning the number of observations, two situations are investigated: $n = 50$ ($n < p$, high-dimensional data) and $n = 150$ ($n > p$). In both cases, the contamination level is set to $\varepsilon = 0.1$. The number of predictors in the final RCS model is limited to the number of latent variables $k = 5$. For the remaining parameters of RCS, the same settings as in the example from Section 8.3 are used, i.e., 15 variables are sequenced in the initial B-RLARS step with 50 bootstrap repetitions, the significance level for MM-regression in the second step is set to $\alpha = 0.3$, and the BIC used as criterion for k-subset regression in the third step. In addition, the simulations are performed with the R package **simFrame** (Alfons 2010, Alfons et al. 2010c), which is a general framework for statistical simulation.

The results from 100 simulation runs are presented in Table 8.4. Averages of certain quantities of interest are thereby computed. The final RCS model is evaluated by the number of groups of low-noise perturbations that are represented by exactly one variable (#target), the number of noise variables (#noise), and the total number of variables (#total). Ideally, the final model would consist of $k = 5$ target predictors—exactly one from each group and no noise variables. Since the success of the procedure depends on the initial B-RLARS step, the initial sequence from this step is evaluated as well. As discussed in the example in Section 8.3, the first part of the sequence may not contain some important predictors. In order to further investigate this issue, the number of groups that are represented by exactly one variable (#target) and the number of noise variables (#noise) are computed for the first k variables in the initial B-RLARS sequence as well. In the complete B-RLARS sequence, as many of the low-noise perturbations as possible should be included. It is essential that all groups occur in the sequence so that it is possible to extract one variable for each group in the remaining steps of the procedure. Therefore, the initial B-RLARS sequence is evaluated using the number of represented groups (#groups) and the number of noise variables (#noise).

8. Robust variable selection

Table 8.4: Average results from 100 simulation runs with contamination level $\varepsilon = 0.1$. For RCS, the number of target groups represented by exactly one variable (#target), the number of noise variables (#noise), and the total number of variables (#total) are shown. For the first k variables of B-RLARS, the number of target groups represented by exactly one variable (#target) and the number of noise variables (#noise) are displayed. The full B-RLARS sequence is evaluated using the number of represented groups (#groups) and the number of noise variables (#noise).

		RCS			First k of B-RLARS		B-RLARS	
n	Scenario	#target	#noise	#total	#target	#noise	#groups	#noise
50	1	4.84	0.06	4.90	3.91	0.43	4.99	5.25
	2	4.79	0.09	4.88	3.82	0.52	5	5.58
	3	4.28	0.68	4.96	3.48	1.02	4.82	7.47
150	1	5	0	5	3.86	0	5	1.19
	2	5	0	5	4.12	0.02	5	1.81
	3	4.89	0.11	5	3.91	0.49	5	4.84

The initial B-RLARS step performs well in this setting if all variables from the groups of low-noise perturbations and no additional noise variables are sequenced.

The simulation results from Table 8.4 indicate that the RCS procedure performs very well. In particular in the case of $n > p$, the results are excellent. Only in some instances for the scenario with contamination in the candidate predictors, the final model does not contain exactly one variable from each group of low-noise perturbations. In these instances, the final model also contains one noise variable, which may be due to the considerably higher number of noise variables in the initial B-RLARS sequence compared to the other scenarios. In the case of $n < p$ (low sample size, high-dimensional data), variable selection is much more difficult, which is also reflected in the simulation results. For all scenarios, the number of noise variables in the initial B-RLARS sequence is much higher than in the case of $n > p$. The RCS procedure still gives excellent results if the data are not contaminated or if contamination is only present in the response. Merely in some cases, the final model consists of less than $k = 5$ predictors or contains a noise variable. But even if the candidate predictors are contaminated as well, the results are very reasonable considering that on average about half of the variables in the initial B-RLARS sequence are noise variables.

Furthermore, the results from the simulations show that the first parts of the B-RLARS sequence may not contain some important variables for data of a certain structure. In all investigated scenarios, the first k variables in the initial B-RLARS sequence often contain more than one variable from the same group of low-noise perturbations, and in some scenarios even noise variables frequently occur.

8.5 Conclusions and discussion

Motivated by a practical application, we developed a strategy for finding a linear regression model that includes only a necessary minimum of key predictor variables to describe the response. The number of explanatory variables thereby was supposed to be smaller than a given boundary, each of them should contain potentially new information, and the resulting model should be highly interpretable. Moreover, the variable selection procedure needed to be robust with respect to possible data inhomogeneities and outliers. The difficulty with these requirements was that the underlying data set is high-dimensional, with much more variables than observations.

Several methods for model selection in high dimensions are available to date, but only a few proposals for robust model selection have been made due to the much higher request of computation time. Our algorithm is based on bootstrapped robust least angle regression (B-RLARS; Khan et al. 2007b), which we apply to find an initial sequence of explanatory variables. In addition to being robust to atypical observations, B-RLARS yields a stable sequence of predictors because of the bootstrap procedure, it is fast to compute, and R code (R Development Core Team 2010) is freely available. Different strategies for further reducing the initial sequence of predictor variables are possible. Since our aim is to extract a small set of highly informative explanatory variables, filtering out the non-significant variables with MM-regression (Yohai 1987, Maronna et al. 2006) seems a suitable approach. MM-regression is used because it is both highly efficient and highly robust. Then all subsets of a given maximum size k of the set of significant variables can be examined to find the optimal regression model. However, using robust regression and resampling methods for this purpose is computationally expensive. Therefore, we suggest using k-subset regression based on least squares (e.g., Furnival and Wilson 1974, Miller 2002, Gatu and Kontoghiorghes 2006), which is robustified by using the weights obtained from another MM-regression model with only the significant explanatory variables. This is a simplification because the weights obtained from MM-regression on the significant variables might not be appropriate for a subset of these variables. For this reason, an alternative procedure based on the root trimmed mean squared error of prediction (RTMSEP) has been proposed as well, which nevertheless is computationally much more demanding. Note that also other procedures for robust variable selection are possible, such as the forward search strategy (see Atkinson and Riani 2002).

In the example of extracting a small set of explanatory variables for quality of life, the suggested strategy succeeded in finding an easy-to-interpret model containing only predictors with potentially new information. The latter was confirmed by a cluster analysis based on robust correlations (see Figure 8.2). Moreover, the resulting model is an excellent fit and performs well with respect to prediction. Simulation results were presented as further indication of the excellent performance of the proposed procedure. Last but not least, our

8. Robust variable selection

procedure also gave meaningful answers to other questions and hypotheses related to the project.

A principal question is whether robust methods are really required for a data set at hand. Usually, inspecting high-dimensional data for possible inhomogeneities or outliers is difficult. For our data set, we used the outlier detection method by Filzmoser et al. (2008), which identified some clearly outlying observations. In the example for quality of life, the weights obtained by MM-regression with the reduced set of predictor variables indicated that outliers still exist in the much lower-dimensional subset of the data. In any case, even if only minor contamination is present, robust model selection can yield more stable results, as it is less sensitive to small changes in the (high-dimensional) data.

Acknowledgements The authors are grateful to the referees for helpful comments and suggestions.

References

D. Adler, C. Gläser, O. Nenadic, J. Oehlschlägel, and W. Zucchini. *ff: Memory-efficient storage of large atomic vectors and arrays on disk and fast access functions*, 2010. URL http://ff.R-Forge.R-project.org. R package version 2.2-1.

J. Aitchison. *The Statistical Analysis of Compositional Data*. Chapman & Hall, London, 1986. ISBN 0-412-28060-4.

J. Aitchison. On criteria for measures of compositional difference. *Mathematical Geology*, 24(4):365–379, 1992.

J. Aitchison, C. Barceló-Vidal, J.A. Martín-Fernández, and V. Pawlowsky-Glahn. Logratio analysis and compositional distance. *Mathematical Geology*, 32(3):271–275, 2000.

H. Akaike. Statistical predictor identification. *Annals of the Institute of Statistical Mathematics*, 22(2):203–217, 1970.

A. Alfons. *simFrame: Simulation framework*, 2010. URL http://CRAN.R-project.org/package=simFrame. R package version 0.3.6.

A. Alfons and S. Kraft. *simPopulation: Simulation of synthetic populations for surveys based on sample data*, 2010. URL http://CRAN.R-project.org/package=simPopulation. R package version 0.2.1.

A. Alfons, M. Templ, P. Filzmoser, S. Kraft, and B. Hulliger. Intermediate report on the data generation mechanism and on the design of the simulation study. AMELI Deliverable 6.1, Department of Statistics and Probability Theory, Vienna University of Technology, 2009. URL http://ameli.surveystatistics.net.

A. Alfons, J. Holzer, and M. Templ. *laeken: Laeken indicators for measuring social cohesion*, 2010a. URL http://CRAN.R-project.org/package=laeken. R package version 0.1.3.

A. Alfons, S. Kraft, M. Templ, and P. Filzmoser. Simulation of synthetic population data for household surveys with application to EU-SILC. Research Report CS-2010-1, Department of Statistics and Probability Theory, Vienna University of Technology, 2010b. URL http://www.statistik.tuwien.ac.at/forschung/CS/CS-2010-1complete.pdf.

REFERENCES

A. Alfons, M. Templ, and P. Filzmoser. An object-oriented framework for statistical simulation: The R package **simFrame**. *Journal of Statistical Software*, 37(3):1–36, 2010c. URL http://www.jstatsoft.org/v37/i03/.

A. Alfons, M. Templ, and P. Filzmoser. Contamination models in the R package **simFrame** for statistical simulation. In S. Aivazian, P. Filzmoser, and Y. Kharin, editors, *Computer Data Analysis and Modeling: Complex Stochastic Data and Systems*, volume 2, pages 178–181, Minsk, 2010d. ISBN 978-985-476-848-9.

A. Alfons, M. Templ, P. Filzmoser, and J. Holzer. A comparison of robust methods for Pareto tail modeling in the case of Laeken indicators. In C. Borgelt, G. González-Rodríguez, W. Trutschnig, M.A. Lubiano, M.A. Gil, P. Grzegorzewski, and O. Hryniewicz, editors, *Combining Soft Computing and Statistical Methods in Data Analysis*, volume 77 of *Advances in Intelligent and Soft Computing*, pages 17–24. Springer, Heidelberg, 2010e. ISBN 978-3-642-14745-6.

A. Alfons, W.E. Baaske, P. Filzmoser, W. Mader, and R. Wieser. Robust variable selection with application to quality of life research. *Statistical Methods & Applications*, 20(1): 65–82, 2011a.

A. Alfons, S. Kraft, M. Templ, and P. Filzmoser. Simulation of close-to-reality population data for household surveys with application to EU-SILC. *Statistical Methods & Applications*, 2011b. URL http://dx.doi.org/10.1007/s10260-011-0163-2. DOI 10.1007/s10260-011-0163-2, to appear.

K. Arnold, J. Gosling, and D. Holmes. *The Java Programming Language*. Prentice Hall, Upper Saddle River, 4th edition, 2005. ISBN 978-0321349804.

A.C. Atkinson and M. Riani. Forward search added-variable t-tests and the effect of masked outliers on model selection. *Biometrika*, 89(4):939–946, 2002.

T. Atkinson, B. Cantillon, E. Marlier, and B. Nolan. *Social Indicators: The EU and Social Inclusion*. Oxford University Press, New York, 2002. ISBN 0-19-925349-8.

W.E. Baaske, P. Filzmoser, W. Mader, and R. Wieser. Agriculture as a success factor for municipalities. In *Jahrbuch der Österreichischen Gesellschaft für Agrarökonomie (ÖGA)*, volume 18, pages 21–30. Facultas Verlag, Vienna, 2009. ISBN 978-3-7089-0432-3.

C. Béguin and B. Hulliger. The BACON-EEM algorithm for multivariate outlier detection in incomplete survey data. *Survey Methodology*, 34(1):91–103, 2008.

J. Beirlant, P. Vynckier, and J.L. Teugels. Tail index estimation, Pareto quantile plots, and regression diagnostics. *Journal of the American Statistical Association*, 31(436):1659–1667, 1996a.

REFERENCES

J. Beirlant, P. Vynckier, and J.L. Teugels. Excess functions and estimation of the extreme-value index. *Bernoulli*, 2(4):293–318, 1996b.

A. Burton, D.G. Altman, P. Royston, and R.L. Holder. The design of simulation studies in medical statistics. *Statistics in Medicine*, 25(24):4279–4292, 2006.

J.M. Chambers. *Software for Data Analysis: Programming with R*. Springer, New York, 2008. ISBN 978-0-387-75935-7.

J.M. Chambers. *Programming with Data*. Springer, New York, 1998. ISBN 0-387-98503-4.

J.M Chambers and T.J. Hastie. *Statistical Models in S*. Chapman & Hall, London, 1992. ISBN 9780412830402.

R. Chambers. Evaluation criteria for statistical editing and imputation. EurEdit Deliverable D3.3, Department of Social Statistics, University of Southhampton, 2001.

R.L. Chambers. Outlier robust finite population estimation. *Journal of the American Statistical Association*, 81(396):1063–1069, 1986.

G. Chauvet and Y. Tillé. A fast algorithm of balanced sampling. *Computational Statistics*, 21(1):53–62, 2006.

X.-H. Chen, A.P. Dempster, and J.S. Liu. Weighted finite population sampling to maximize entropy. *Biometrika*, 81(3):457–469, 1994.

H. Choi and N.M. Kiefer. Improving robust model selection tests for dynamic models. *Econometrics Journal*, 13(2):177–204, 2010.

G.P. Clarke. Microsimulation: An introduction. In G.P. Clarke, editor, *Microsimulation for Urban and Regional Policy Analysis*. Pion, London, 1996.

W.G. Cochran. *Sampling Techniques*. John Wiley & Sons, New York, 3rd edition, 1977. ISBN 0-471-16240-X.

C. Croux and C. Dehon. Influence functions of the Spearman and Kendall correlation measures. *Statistical Methods & Applications*, 19(4):497–515, 2010.

C. Croux, P. Filzmoser, G. Pison, and P.J. Rousseeuw. Fitting multiplicative models by robust alternating regressions. *Statistics and Computing*, 13(1):23–36, 2003.

C. Croux, G. Dhaene, and D. Hoorelbeke. Robust standard errors for robust estimators. Discussion Papers Series 03.16, KU Leuven, 2008.

J.-C. Deville and Y. Tillé. Efficient balanced sampling: The cube method. *Biometrika*, 91(4):893–912, 2004.

REFERENCES

J.-C. Deville and Y. Tillé. Unequal probability sampling without replacement through a splitting method. *Biometrika*, 85(1):89–101, 1998.

J.-C. Deville, C.-E. Särndal, and O. Sautory. Generalized raking procedures in survey sampling. *Journal of the American Statistical Association*, 88(423):1013–1020, 1993.

E. Diener, E.M. Suh, R.E. Lucas, and H.L. Smith. Subjective well-being: Three decades of progress. *Psycholigical Bulletin*, 125(2):276–302, 1999.

J. Domingo-Ferrer and J.M. Mateo-Sanz. Practical data-oriented microaggregation for statistical disclosure control. *IEEE Transactions on Knowledge and Data Engineering*, 14(1): 189–201, 2002.

D.L. Donoho and P.J. Huber. The notion of breakdown point. In P.J. Bickel, K. Doksum, and J.L. Hodges, Jr., editors, *A Festschrift for Erich L. Lehmann*. Wadsworth, Belmont, 1983.

J. Drechsler, S. Bender, and S. Rässler. Comparing fully and partially synthetic datasets for statistical disclosure control in the German IAB Establishment Panel. *Transactions on Data Privacy*, 1(3):105–130, 2008.

G.T. Duncan and D. Lambert. Disclosure-limited data dissemination. *Journal of the American Statistical Association*, 81(393):10–28, 1986.

D.J. Dupuis and S. Morgenthaler. Robust weighted likelihood estimators with an application to bivariate extreme value problems. *The Canadian Journal of Statistics*, 30(1):17–36, 2002.

D.J. Dupuis and M.-P. Victoria-Feser. A robust prediction error criterion for Pareto modelling of upper tails. *The Canadian Journal of Statistics*, 34(4):639–658, 2006.

B. Efron, T. Hastie, I. Johnstone, and R. Tibshirani. Least angle regression. *The Annals of Statistics*, 32(2):407–499, 2004.

J.J. Egozcue, V. Pawlowsky-Glahn, G. Mateu-Figueras, and C. Barceló-Vidal. Isometric logratio transformations for compositional data analysis. *Mathematical Geology*, 35(3): 279–300, 2003.

E.A.H. Elamir and C.J. Skinner. Record level measures of disclosure risk for survey microdata. *Journal of Official Statistics*, 22(3):525–539, 2006.

P. Embrechts, G. Klüppelberg, and T. Mikosch. *Modelling Extremal Events for Insurance and Finance*. Springer, New York, 1997. ISBN 3-540-60931-8.

REFERENCES

EU-SILC. Common cross-sectional EU indicators based on EU-SILC; the gender pay gap. EU-SILC 131-rev/04, Working group on Statistics on Income and Living Conditions (EU-SILC), Eurostat, Luxembourg, 2004.

Eurostat. Description of target variables: Cross-sectional and longitudinal. EU-SILC 065/04, Eurostat, Luxembourg, 2004.

B.S. Everitt and G. Dunn. *Applied Multivariate Data Analysis*. Arnold, London, 2nd edition, 2001. ISBN 0-340-54529-1.

S.E. Fienberg, U.E. Makov, and A.P. Sanil. A bayesian approach to data disclosure: Optimal intruder behavior for continuous data. *Journal of Official Statistics*, 13(1):75–89, 1997.

P. Filzmoser, R. Maronna, and M. Werner. Outlier identification in high dimensions. *Computational Statistics & Data Analysis*, 52(3):1694–1711, 2008.

M. Fowler. **UML Distilled:** *A Brief Guide to the Standard Object Modeling Language*. Addison-Wesley, 3rd edition, 2003. ISBN 978-0-321-19368-1.

L. Franconi and S. Polettini. Individual risk estimation in μ-ARGUS: A review. In J. Domingo-Ferrer and V. Torra, editors, *Privacy in Statistical Databases*, volume 3050 of *Lecture Notes in Computer Science*, pages 262–272. Springer, Heidelberg, 2004. ISBN 978-3-540-22118-2.

G. Furnival and R. Wilson. Regression by leaps and bounds. *Technometrics*, 16(4):499–511, 1974.

C. Gatu and E.J. Kontoghiorghes. Branch-and-bound algorithms for computing the best-subset regression models. *Journal of Computational and Graphical Statistics*, 15(1):139–156, 2006.

A. Genz and F. Bretz. *Computation of Multivariate Normal and t Probabilities*, volume 195 of *Lecture Notes in Statistics*. Springer, New York, 2009. ISBN 978-3-642-01688-2.

A. Genz, F. Bretz, T. Miwa, X. Mi, F. Leisch, F. Scheipl, and T. Hothorn. **mvtnorm:** *Multivariate normal and t distributions*, 2010. URL http://CRAN.R-project.org/package=mvtnorm. R package version 0.9-92.

M. Haahr. random.org: True random number service, 2010. URL http://www.random.org. Accessed October 20, 2010.

J. Hájek. Asymptotic theory of rejective sampling with varying probabilities from a finite population. *Annals of Mathematical Statistics*, 35(4):1491–1523, 1964.

REFERENCES

F.R. Hampel. A general qualitative definition of robustness. *Annals of Mathematical Statistics*, 42(6):1887–1896, 1971.

F.R. Hampel. The influence curve and its role in robust estimation. *Journal of the American Statistical Association*, 69(346):383–393, 1974.

T. Hastie, R. Tibshirani, and J. Friedman. *The Elements of Statistical Learning*. Springer, New York, 2nd edition, 2009. ISBN 978-0-387-84857-0.

B.M. Hill. A simple general approach to inference about the tail of a distribution. *The Annals of Statistics*, 3(5):1163–1174, 1975.

J. Holzer. Robust methods for the estimation of selected Laeken indicators. Master's thesis, Department of Statistics and Probability Theory, Vienna University of Technology, Vienna, Austria, 2009.

D.G. Horvitz and D.J. Thompson. A generalization of sampling without replacement from a finite universe. *Journal of the American Statistical Association*, 47(260):663–685, 1952.

K. Hron, M. Templ, and P. Filzmoser. Imputation of missing values for compositional data using classical and robust methods. *Computational Statistics & Data Analysis*, 54(12): 3095–3107, 2010.

B. Hulliger and T. Schoch. Robustification of the quintile share ratio. New Techniques and Technologies for Statistics, Brussels, 2009a.

B. Hulliger and T. Schoch. Robust multivariate imputation with survey data. 57th Session of the International Statistical Institute, Durban, 2009b.

M.E. Johnson. *Multivariate Statistical Simulation*. John Wiley & Sons, New York, 1987. ISBN 0-471-82290-6.

O. Jones, R. Maillardet, and A. Robinson. *Introduction to Scientific Programming and Scientific Simulation Using R*. Chapman & Hall/CRC, Boca Raton, 2009. ISBN 978-1-4200-6872-6.

M.G. Kendall and A. Stuart. *The Advanced Theory of Statistics*, volume 2. Charles Griffin & Co. Ltd., London, 2nd edition, 1967.

J.A. Khan, S. Van Aelst, and R.H. Zamar. Building a robust linear model with forward selection and stepwise procedures. *Computational Statistics & Data Analysis*, 52(1):239–248, 2007a.

J.A. Khan, S. Van Aelst, and R.H. Zamar. Robust linear model selection based on least angle regression. *Journal of the American Statistical Association*, 102(480):1289–1299, 2007b.

REFERENCES

C. Kleiber and S. Kotz. *Statistical Size Distributions in Economics and Actuarial Sciences*. John Wiley & Sons, Hoboken, 2003. ISBN 0-471-15064-9.

S. Kraft. Simulation of a population for the European Income and Living Conditions survey. Master's thesis, Department of Statistics and Probability Theory, Vienna University of Technology, Vienna, Austria, 2009.

P. L'Ecuyer and J. Leydold. **rstream**: Streams of random numbers for stochastic simulation. *R News*, 5(2):16–20, 2005. URL http://CRAN.R-project.org/doc/Rnews/.

P. L'Ecuyer, R. Simard, E.J. Chen, and W.D. Kelton. An object-oriented random-number package with many long streams and substreams. *Operations Research*, 50(6):1073–1075, 2002.

F. Leisch. **Sweave**: Dynamic generation of statistical reports using literate data analysis. In W. Härdle and B. Rönz, editors, *Compstat 2002 - Proceedings in Computational Statistics*, pages 575–580, Heidelberg, 2002a. Physica Verlag. ISBN 3-7908-1517-9.

F. Leisch. **Sweave**, part I: Mixing R and LATEX. *R News*, 2(3):28–31, 2002b.

F. Leisch. **Sweave**, part II: Package vignettes. *R News*, 3(2):21–24, 2003.

J. Leydold. *rstream: Streams of random numbers*, 2010. URL http://statistik.wu-wien.ac.at/arvag/. R package version 1.2.5.

N. Li. *rsprng: R interface to **SPRNG** (Scalable Parallel Random Number Generators)*, 2010. URL http://CRAN.R-project.org/package=rsprng. R package version 1.0.

R.J.A. Little. Statistical analysis of masked data. *Journal of Official Statistics*, 9(2):407–426, 1993.

R.J.A. Little and D.B. Rubin. *Statistical Analysis with Missing Data*. John Wiley & Sons, New York, 2nd edition, 2002. ISBN 0-471-18386-5.

M.O. Lorenz. Methods of measuring the concentration of wealth. *Publications of the American Statistical Association*, 9(70):209–219, 1905.

T. Lumley and A. Miller. *leaps: Regression subset selection*, 2009. URL http://CRAN.R-project.org/package=leaps. R package version 2.9.

C.L. Mallows. Some comments on C_p. *Technometrics*, 15(4):661–675, 1973.

R. Maronna, D. Martin, and V. Yohai. *Robust Statistics*. John Wiley & Sons, Chichester, 2006. ISBN 978-0-470-01092-1.

REFERENCES

R.A. Maronna and R.H. Zamar. Robust estimates of location and dispersion for high-dimensional datasets. *Technometrics*, 44(4):307–317, 2002.

M. Mascagni and A. Srinivasan. Algorithm 806: **SPRNG**: A scalable library for pseudo-random number generation. *ACM Transactions on Mathematical Software*, 26(3):436–461, 2000.

G. Mateu-Figueras, V. Pawlowsky-Glahn, and J.J. Egozcue. The normal distribution in some constrained sample spaces, 2008. URL http://arxiv.org/abs/0802.2643.

M. Matsumoto and T. Nishimura. Mersenne Twister: A 623-dimensionally equidistributed uniform pseudo-random number generator. *ACM Transactions on Modeling and Computer Simulation*, 8(1):3–30, 1998.

L. McCann and R.E. Welsch. Robust variable selection using least angle regression and elemental set sampling. *Computational Statistics & Data Analysis*, 52(1):249–257, 2007.

D. Meyer, A. Zeileis, and K. Hornik. The strucplot framework: Visualizing multi-way contingency tables with **vcd**. *Journal of Statistical Software*, 17(3):1–48, 2006.

D. Meyer, A. Zeileis, and K. Hornik. **vcd**: *Visualizing categorical data*, 2010. URL http://CRAN.R-project.org/package=vcd. R package version 1.2-9.

H. Midzuno. On the sampling system with probability proportional to sum of size. *Annals of the Institute of Statistical Mathematics*, 3(2):99–107, 1952.

A. Miller. *Subset Selection in Regression*. Chapman & Hall/CRC, Boca Raton, 2nd edition, 2002. ISBN 1-58488-171-2.

B.J.T. Morgan. *Elements of Simulation*. Chapman & Hall, London, 1984. ISBN 0-412-24590-6.

S. Müller and A.H. Welsh. Outlier robust model selection in linear regression. *Journal of the American Statistical Association*, 100(472):1297–1310, 2005.

R. Münnich and J. Schürle. On the simulation of complex universes in the case of applying the German Microcensus. DACSEIS research paper series No. 4, University of Tübingen, 2003. URL http://w210.ub.uni-tuebingen.de/volltexte/2003/979/.

R. Münnich, W. Bihler, J. Bjørnstad, Z. Li-Chun, A. Davidson, S. Sardy, A. Haslinger, P. Knotterus, S. Laaksonen, D. Ohly, J. Schürle, R. Wiegert, U. Oetliker, J.-P. Renfer, A. Quatember, C. Skinner, and Y. Berger. Data quality in complex surveys. DACSEIS Deliverable D1.1, University of Tübingen, 2003a. URL http://www.dacseis.de.

REFERENCES

R. Münnich, J. Schürle, W. Bihler, H.-J. Boonstra, P. Knotterus, N. Nieuwenbroek, A. Haslinger, S. Laaksonen, D. Eckmair, A. Quatember, H. Wagner, J.-P. Renfer, U. Oetliker, and R. Wiegert. Monte Carlo simulation study of European surveys. DACSEIS Deliverables D3.1 and D3.2, University of Tübingen, 2003b. URL http://www.dacseis.de.

T.E. Raghunathan, J.P. Reiter, and D.B. Rubin. Multiple imputation for statistical disclosure limitation. *Journal of Official Statistics*, 19(1), 2003.

J.N.K. Rao. *Small Area Estimation*. John Wiley & Sons, Hoboken, 2003. ISBN 978-0-471-41374-5.

R Development Core Team. *R: A language and environment for statistical computing*. R Foundation for Statistical Computing, Vienna, Austria, 2010. URL http://www.R-project.org. ISBN 3-900051-07-0.

J.P. Reiter. Using multiple imputation to integrate and disseminate confidential microdata. *International Statistical Review*, 77(2):179–195, 2009.

J.P. Reiter and R. Mitra. Estimating risks of identification disclosure in partially synthetic data. *Journal of Privacy and Confidentiality*, 1(1):99–110, 2009.

R. Renwick, I. Brown, and D. Raphael. Quality of life: Linking conceptual approach to service provision. *Journal on Developmental Disabilities*, 3(2):32–44, 1994.

M. Riani and A.C. Atkinson. Robust model selection with flexible trimming. *Computational Statistics & Data Analysis*, 54(12):3300–3312, 2010.

Y. Rinott and N. Shlomo. A generalized negative binomial smoothing model for sample disclosure risk estimation. In J. Domingo-Ferrer and L. Franconi, editors, *Privacy in Statistical Databases*, volume 4302 of *Lecture Notes in Computer Science*, pages 82–93. Springer, Heidelberg, 2006. ISBN 978-3-540-49330-3.

B.D. Ripley. *Stochastic Simulation*. John Wiley & Sons, New York, 1987. ISBN 0-471-81884-4.

E. Ronchetti and R.G. Staudte. A robust version of Mallows's C_p. *Journal of the American Statistical Association*, 89(426):550–559, 1994.

E. Ronchetti, C. Field, and W. Blanchard. Robust linear model selection by cross-validation. *Journal of the American Statistical Association*, 92(439):1017–1023, 1997.

A.J. Rossini, L. Tierney, and N. Li. Simple parallel statistical computing in R. *Journal of Computational and Graphical Statistics*, 16(2):399–420, 2007.

REFERENCES

P.J. Rousseeuw and A.M. Leroy. *Robust Regression and Outlier Detection*. John Wiley & Sons, New York, 1987. ISBN 0-471-48855-0.

P.J. Rousseeuw and K. Van Driessen. A fast algorithm for the minimum covariance determinant estimator. *Technometrics*, 41(3):212–223, 1999.

P.J. Rousseeuw, C. Croux, V. Todorov, A. Ruckstuhl, M. Salibian-Barrera, T. Verbeke, and M. Maechler. **robustbase**: *Basic robust statistics*, 2009. URL http://CRAN.R-project.org/package=robustbase. R package version 0.5-0-1.

D. Rubin. Inference and missing data. *Biometrika*, 63(3):581–592, 1976.

D.B. Rubin. Discussion: Statistical disclosure limitation. *Journal of Official Statistics*, 9(2): 461–468, 1993.

M. Salibian-Barrera and S. Van Aelst. Robust model selection using fast and robust bootstrap. *Computational Statistics & Data Analysis*, 52(12):5121–5135, 2008.

M. Salibian-Barrera and R.H. Zamar. Bootstrapping robust estimates of regression. *The Annals of Statistics*, 30(2):556–582, 2002.

P. Samerati and L. Sweeney. Protecting privacy when disclosing information: k-anonymity and its enforcement through generalization and suppression. Technical Report SRI CSL 98-04, SRI International, 1998.

M.R. Sampford. On sampling without replacement with unequal probabilities of selection. *Biometrika*, 54(2):499–513, 1967.

D. Sarkar. *Lattice: Multivariate Data Visualization with R*. Springer, New York, 2008. ISBN 978-0-387-75968-5.

D. Sarkar. **lattice**: *Lattice graphics*, 2010. URL http://CRAN.R-project.org/package=lattice. R package version 0.19-13.

C.-E. Särndal, B. Swensson, and J. Wretman. *Model Assisted Survey Sampling*. Springer, New York, 2003. ISBN 0-387-40620-4.

M. Schmidberger, M. Morgan, D. Eddelbuettel, H. Yu, L. Tierney, and U. Mansmann. State of the art in parallel computing with R. *Journal of Statistical Software*, 31(1):1–27, 2009. URL http://www.jstatsoft.org/v31/i01.

G. Schwarz. Estimating the dimension of a model. *The Annals of Statistics*, 6(2):461–464, 1978.

REFERENCES

H. Sevcikova and T. Rossini. **rlecuyer**: *R interface to RNG with multiple streams*, 2009. URL http://CRAN.R-project.org/package=rlecuyer. R package version 0.3-1.

J.S. Simonoff. *Analyzing Categorical Data*. Springer, New York, 2003. ISBN 0-387-00749-0.

L. Sweeney. k-anonymity: A model for protecting privacy. *International Journal on Uncertainty, Fuzziness and Knowledge-based Systems*, 10(5):557–570, 2002.

M. Templ and A. Alfons. Disclosure risk of synthetic population data with application in the case of EU-SILC. In J. Domingo-Ferrer and E. Magkos, editors, *Privacy in Statistical Databases*, volume 6344 of *Lecture Notes in Computer Science*, pages 174–186. Springer, Heidelberg, 2010. ISBN 978-3-642-15837-7.

M. Templ and P. Filzmoser. Visualization of missing values using the R-package **VIM**. Research Report CS-2008-1, Department of Statistics and Probability Theory, Vienna University of Technology, 2008. URL http://www.statistik.tuwien.ac.at/forschung/CS/CS-2008-1complete.pdf.

M. Templ and B. Meindl. Robust statistics meets SDC: New disclosure risk measures for continuous microdata masking. In J. Domingo-Ferrer and Y. Saygin, editors, *Privacy in Statistical Databases*, volume 5262 of *Lecture Notes in Computer Science*, pages 113–126. Springer, Heidelberg, 2008. ISBN 978-3-540-87470-6.

M. Templ, P. Filzmoser, and K. Hron. Robust imputation of missing values in compositional data using the R-package **robCompositions**. New Techniques and Technologies for Statistics, Brussels, 2009.

M. Templ, A. Alfons, and A. Kowarik. *VIM: Visualization and imputation of missing values*, 2010a. URL http://CRAN.R-project.org/package=VIM. R package version 1.4.2.

M. Templ, K. Hron, and P. Filzmoser. **robCompositions**: *Robust estimation for compositional data.*, 2010b. URL http://CRAN.R-project.org/package=robCompositions. R package version 1.4.3.

G. Terrell. Linear density estimates. In *Proceedings of the Statistical Computing Section*, pages 297–302. American Statistical Association, 1990.

C. Tichbon and P. Newton. Life is do-able: Quality of life development in a supportive small group setting. Occasional Paper Series 2, Mental Health Foundation of New Zealand, 2002.

L. Tierney, A.J. Rossini, N. Li, and H. Sevcikova. **snow**: *Simple network of workstations*, 2008. URL http://CRAN.R-project.org/package=snow. R package version 0.3-3.

REFERENCES

L. Tierney, A.J. Rossini, and N. Li. **snow**: A parallel computing framework for the R system. *International Journal of Parallel Programming*, 37(1):78–90, 2009.

Y. Tillé. *Sampling Algorithms*. Springer, New York, 2006. ISBN 0-387-30814-8.

Y. Tillé. An elimination procedure of unequal probability sampling without replacement. *Biometrika*, 83(1):238–241, 1996.

Y. Tillé and A. Matei. **sampling**: *Survey sampling*, 2009. URL http://CRAN.R-project.org/package=sampling. R package version 2.3.

V. Todorov. **rrcov**: *Scalable robust estimators with high breakdown point*, 2010. URL http://CRAN.R-project.org/package=rrcov. R package version 1.1-00.

V. Todorov and P. Filzmoser. An object-oriented framework for robust multivariate analysis. *Journal of Statistical Software*, 32(3):1–47, 2009. URL http://www.jstatsoft.org/v32/i03.

O. Troyanskaya, M. Cantor, G. Sherlock, P. Brown, T. Hastie, R. Tibshirani, D. Botstein, and R.B. Altman. Missing value estimation methods for DNA microarrays. *Bioinformatics*, 17(6):520–525, 2001.

S. Urbanek. **multicore**: *Parallel processing of R code on machines with multiple cores or CPUs*, 2009. URL http://www.RForge.net/multicore/. R package version 0.1-3.

S. Van Aelst, R. Welsch, and R.H. Zamar, editors. Special issue on variable selection and robust procedures. *Computational Statistics & Data Analysis*, 54(12), 2010.

P. Van Kerm. Extreme incomes and the estimation of poverty and inequality indicators from EU-SILC. IRISS Working Paper Series 2007-01, CEPS/INSTEAD, 2007.

B. Vandewalle, J. Beirlant, A. Christmann, and M. Hubert. A robust estimator for the tail index of Pareto-type distributions. *Computational Statistics & Data Analysis*, 51(12): 6252–6268, 2007.

K. Varmuza and P. Filzmoser. *Introduction to Multivariate Statistical Analysis in Chemometrics*. CRC Press, Boca Raton, 2009. ISBN 978-0-470-98581-6.

A.J. Walker. An efficient method for generating discrete random variables with general distributions. *ACM Transactions on Mathematical Software*, 3(3):253–256, 1977.

S. Weisberg. *Applied Linear Regression*. John Wiley & Sons, Hoboken, 3rd edition, 2005. ISBN 0-471-66379-4.

REFERENCES

J.W. Wisnowski, J.R. Simpson, D.C. Montgomery, and G.C. Runger. Resampling methods for variable selection in robust regression. *Computational Statistics & Data Analysis*, 43(3):341–355, 2003.

V.J. Yohai. High breakdown-point and high efficiency robust estimates for regression. *The Annals of Statistics*, 15(20):642–656, 1987.

Index

AMELI, 1, 2
ARPR, *see* at-risk-of-poverty rate
ARPT, *see* at-risk-of-poverty threshold
at-risk-of-poverty rate, 3
at-risk-of-poverty threshold, 3

box plot, 42, 94
breakdown point, 6
    asymptotic, 6
    finite-sample, 6

calibration, 13
CAR, *see* contaminated at random
CCAR, *see* contaminated completely at random
CDF, *see* cumulative distribution function
class, 23
    diagram, 26
    inheritance, 23
    virtual, 23
complete linkage clustering, 133
contaminated at random, 14, 75
contaminated completely at random, 14, 75
contamination, 5, 34, 74
    in simulation studies, 13, 34, 73
contamination level, 6, 34, 74
contingency coefficient, 97
cross validation, 131
cumulative distribution function, 94
    empirical, 94

DAR, *see* distributed at random
DCAR, *see* distributed completely at random
dendrogram, 133

disclosure risk
    survey data, 110
    synthetic population data, 113
dissimilarity matrix, 133
distributed at random, 14, 35
distributed completely at random, 14, 35

embarrassingly parallel, 17, 42
equivalized disposable income, 3
ErfolgsVision, 1, 4, 128
EU-SILC, 2, 44, 58, 90, 108, 119

generalized Pareto distribution, 86
generic function, 23
Gini coefficient, 4, 44, 58, 120
GPD, *see* generalized Pareto distribution

influence function, 7

k-subset regression, 129
    robust, 131
kernel density plot, 42

laeken, 44, 58, 120
Laeken indicators, 2, 119–126
LARS, *see* least angle regression
learning curve, 136
least angle regression, 128
    bootstrapped robust, 129
    robust, 128
log-transformation, 88
Lorenz curve, 4

MAR, *see* missing at random
MCAR, *see* missing completely at random

INDEX

MCD, *see* minimum covariance determinant estimator
method, 23
    accessor, 23, 28
    mutator, 23, 28
    signature, 24
microsimulation, 80
minimum covariance determinant estimator, 7, 75, 133
missing at random, 15, 37
missing completely at random, 14, 37
missing not at random, 15, 37
missing value rate, 37
MM-regression, 129, 130
MNAR, *see* missing not at random
mosaic plot, 92
multiple imputation
    fully synthetic microdata, 80, 106
    partially synthetic microdata, 106

OAR, *see* outlying at random
object-oriented programming, 23
OCAR, *see* outlying completely at random
outlier, 5
    nonrepresentative, 8
    representative, 8
outlier detection, 75
outlying at random, 13, 75
outlying completely at random, 13, 75

package vignette, 56, 72, 89
parallel computing, 17, 42
Pareto distribution, 44, 58, 121
    Hill estimator, 44, 58, 122
    partial density component estimator, 44, 58, 123
    weighted maximum likelihood estimator, 122
poststratification, 13

QSR, *see* quintile share ratio
quality of life, 4, 134
quintile share ratio, 4, 120

random error terms, 87
random number generator, 9
    parallel, 18, 42
regression model
    linear, 87
    logistic, 86
    multinomial, 83, 85
    two-step, 86
RNG, *see* random number generator
robust R^2, 136
robust statistics, 5
robust variable selection, 127–144
    context-sensitive, 129
root trimmed mean squared error of prediction, 131
RTMSEP, *see* root trimmed mean squared error of prediction

sample weights, 13
sampling, 12
    balanced, 13
    cluster, 12
    multi-stage, 13
    simple random, 12
    stratified, 12
    unequal probability, 13
signal-to-noise ratio, 140
simFrame, 21–77, 97, 124, 141
simPopulation, 44, 58, 89, 108, 124
simulation, 9
    contamination, 10, 13, 34, 73
    design-based, 9, 10, 30, 44, 57
    missing values, 10, 16, 36
    model-based, 9, 29, 46
    Monte Carlo, 9, 17

INDEX

parallel computing, 49, 69
slot, 23
success factors, 4
synthetic population data, 79–117
    categorical variables, 83, 108
    components, 88, 108
    continuous variables, 85, 108
    data confidentiality, 111
    disclosure scenarios, 112
    household structure, 83, 107

tail modeling, 44, 58, 86, 121
trimming, 87

UML, *see* Unified Modeling Language
Unified Modeling Language, 26

Die VDM Verlagsservicegesellschaft sucht für wissenschaftliche Verlage abgeschlossene und herausragende

Dissertationen, Habilitationen, Diplomarbeiten, Master Theses, Magisterarbeiten usw.

für die kostenlose Publikation als Fachbuch.

Sie verfügen über eine Arbeit, die hohen inhaltlichen und formalen Ansprüchen genügt, und haben Interesse an einer honorarvergüteten Publikation?

Dann senden Sie bitte erste Informationen über sich und Ihre Arbeit per Email an *info@vdm-vsg.de*.

Sie erhalten kurzfristig unser Feedback!

VDM Verlagsservicegesellschaft mbH
Dudweiler Landstr. 99   Telefon +49 681 3720 174
D - 66123 Saarbrücken   Fax   +49 681 3720 1749
www.vdm-vsg.de

Die VDM Verlagsservicegesellschaft mbH vertritt

Printed by Books on Demand GmbH, Norderstedt / Germany